银色锦囊

巴尔塔沙·葛拉西安/著
秦传安/译
卜桦/插图

世界上一半人在嘲笑另一半人，而他们全都是傻瓜

ONE HALF OF THE WORLD LAUGHS AT THE OTHER, AND FOOLS ARE THEY ALL.

是万事皆好，还是万事皆糟，这要取决于你问谁。一个人求之唯恐不得的东西，另一个人害之唯恐不及。依据自己的观念来调整每件事情的人，是一头令人无法忍受的蠢驴。卓越并不依赖于一个人的喜好。有多少人，就有多少种口味，而且各不相同。任何缺点总会有人喜欢。如果某件东西不讨某人喜欢，我们也大可不必灰心丧气，因为总会有别人赏识；如果有人欢呼喝彩我们也用不着沾沾自喜，因为肯定会有其他人来横加责难。对于赞美颂扬，真正的检验标准是声望卓著的人士和本领域专家的首肯。你应该力求不受任何一种观点、任何一种时尚、任何一个世纪的影响。

Everything is good or everything is bad according to who you ask. What one pursues another persecutes. He is an insufferable ass who would regulate everything according to his ideas. Excellences do not depend on a single person's pleasure. So many people, so many tastes, all different. There is no defect that is not affected by some. We need not lose heart if something does not please someone, for others will appreciate it; nor need their applause turn our head, for there will surely be others to condemn it. The real test of praise is the approval of renowned people and of experts in the field. You should aim to be independent of any one opinion, of any one fashion, of any one century.

要能容得下大块的好运

在智慧的身体里，颇为重要的器官是一个大胃,因为大容量意味着大块头。大块的好运，对于能够消化更大块头的胃来说,一点也不为难。撑坏一个胃的东西,对另一个胃来说可能还不饱。有许多胃因其消化力不强而麻烦不断，这要归因于它们的小容量，无论是天生还是后天训练,都不是用来做大事的。它们的机能变质了，不应得的荣誉让它们晕头转向、目眩神迷——高位之上有大风险。它们没有找到合适的位置，因为好运在它们那里也没找到合适的位置。因此,一个天才人物应该让人看到：他有更大的空间,即便是更大事业也能容得下;尤其是要避免让人看到小肚鸡肠的迹象。

BE ABLE TO STOMACH BIG SLICES OF LUCK.

In the body of wisdom not the least important organ is a big stomach, for great capacity implies great parts. Big bits of luck do not embarrass one who can digest still bigger ones. What is a surfeit for one may be hunger for another. Many are troubled as it were with weak digestion, owing to their small capacity being neither born nor trained for great employment. Their actions turn sour, and the fumes that arise from their undeserved honors turn their head and make them dizzy — a great risk in high positions. They do not find their proper place, for luck finds no proper place in them. A person of talent therefore should show that he has more room for even greater enterprises, and above all avoid showing signs of a little heart.

让每个人保持尊严

虽然你并不是国王，但要让你的一言一行都能和君王比肩，一举一动都能向高贵看齐。举止矜持威严，思想崇高深邃，在所有事情上都要像个国王，即使不能表现在权力上，至少也要表现在美德上。真正的王者，在于他的清廉公正，他不需要羡慕伟大，他本人就可以做伟大的楷模。尤其是王座周围的那些人，他们应该志存高远，更应该分享君王的品格，而不仅仅是参与皇家的典仪——分享真正的尊严，忽略其瑕疵。

LET EACH KEEP UP HIS DIGNITY.

Let each deed of a person in its degree, though he be not a king, be worthy of a prince and let his action be princely within due limits. Sublime in action, lofty in thought, in all things like a king, at least in merit if not in might. For true kingship lies in spotless rectitude, and he need not envy greatness who can serve as a model of it. Especially should those near the throne aim at true superiority, and prefer to share the true qualities of royalty rather than take parts in its mere ceremonies — yet without affecting its imperfections but sharing in its true dignity.

www.buhua.com
2004

要懂得不同的职业需要什么

不同的职业，需要不同的才能。要理解这些需要，就要竭尽我们的注意力，调动敏锐的洞察力。有的职业需要勇气，有的则需要机智。那些仅仅需要正直的工作，是最容易的工作，而更难的工作是那些需要心灵手巧的工作。前者所必需的一切就是品格，而后者所必需的一切，一个人的专注和热情恐怕还不够。统治人是件很麻烦的事情，统治傻瓜和笨蛋就更困难了——与那些没有判断力的人在一起工作需要双倍的判断力。当一项职务用固定的工作时间和一成不变的例行公事把一个人占据了的时候，是令人无法忍受的。更好的工作就是让你自由自在地自行其是的、既千变万化又富有意义的工作，因为变化可以让头脑焕然一新。最受人尊敬的工作是那些最少依赖他人的工作。最差劲的工作，就是那些让我们无论是眼下还是往后都烦恼不断的工作。

GET TO KNOW WHAT IS NEEDED IN DIFFERENT OCCUPATIONS.

Different qualities are required. To know which is needed taxes attention and calls for masterly discernment. Some demand courage, others tact. Those that merely require rectitude are the easiest, the more difficult are those requiring cleverness. For the former all that is necessary is character, for the latter all of one's attention and zeal may not suffice. It is a troublesome business to rule people, still more fools or blockheads — twice as much sense is needed with those who have none. It is intolerable when an office engrosses someone with fixed hours and a settled routine. Those are better that leave him free to follow his own devices, combining variety with importance, for the change refreshes the mind. The most respected jobs are those that have least, or most distant, dependence on others. The worst are those that worry us both here and hereafter.

不要招人厌烦

沉迷于同一种行为或话题，容易令人厌烦。简洁令人愉快，收获也更多——其因简略虽有所失，亦必因谦恭而有所得。好事，因简短而倍加其好。提纲挈领比一大堆混乱琐碎的材料更加有效。一个众所周知的事实是：多嘴多舌之徒，很少有聪明人，不管他操心的是身边琐事，还是肩负了多大的职责。有的人，其所扮演的角色与其说是位于中心的装饰品，还不如说是绊手绊脚的障碍物，不过是块无用的废料，挡住了每个人的路。智者避免招人厌烦，尤其不会去招惹大人物——他们的时间都被占满了，打扰他们当中的任何一位，比打扰其他所有人都要糟糕。善言者要言不繁。

DO NOT BE A BORE.

The person obsessed with one activity or one topic is apt to be tiresome. Brevity is flattering and gets more accomplished — it gains by courtesy what it loses by curtness. Good things, when short, are twice as good. The quintessence of the matter is more effective than a big mishmash of details. It is a well-known truth that talkative person rarely is wise, whether in dealing with things at hand or how they function. There are people who serve more as stumbling blocks than centerpieces, useless lumber in everyone's way. The wise avoid being bores, especially to the great — who are fully occupied; it is worse to disturb one of them than all the rest. Well said is soon said.

www.buhua.com
2004

不要炫耀你的身份地位

对自己的身份地位自吹自擂，比狂妄自负更令人讨厌。摆出一副大人物的派头，难免要遭人侧目——想必还会招来足够多的妒嫉。你越是想要寻求别人的尊敬，你所得到的尊敬就越少，因为尊敬不尊敬，还要取决于他人对你的看法。你不能主动索取，而必须以自己的言行赢得并接受他人的尊敬。身居高位，需要运用足够的权威——不如此，则不能完全胜任。因此，要保持足够的尊严，以担负其公共职责。不要强求他人的尊敬，而要努力创造赢得尊敬的条件。那些持守职务尊严的人，往往显得很谦逊，表示自己不敢承当太多的尊敬。如果你希望受到重视，那也应该是因为你的才能，而不是因为任何凭运气得来的东西。即便是贵为君王，他们也更愿意凭藉个人能力而受到尊崇，而不是凭藉他们的身份地位。

DO NOT PARADE YOUR POSITION.

To boast about your position is more offensive than personal vanity. To pose as an important person is to be hated — you should surely have had enough envy. The more you seek esteem the less you obtain it, for it depends on the opinion of others. You cannot take it, but must earn and receive it from others. Great positions require exercising a sufficient amount of authority — without it they cannot be adequately filled. Preserve therefore enough dignity to carry on the duties of the office. Do not enforce respect, but try to create it. Those who insist on the dignity of their office, show they have not deserved it, and that it is too much for them. If you wish to be valued, be valued for your talents, not for anything obtained by chance. Even kings prefer to be honored for their personal qualifications rather than for their station.

www.buhua.com
2004

不要自鸣得意

既不要总是对自己不满，这是缺乏勇气的表现；也不要老是自鸣得意，这是十足的愚蠢。自满多半出于无知，那是一种快乐幸福的无知，即使没让你身败名裂，也不会带来任何好处。一个人无法达到别人所臻至的完美，就会对自己任何平庸的才能沾沾自喜。怀疑是明智的，甚至是有益的，它可以规避灾祸，万一灾祸来临，它也可以为我们提供慰藉。灾祸不会让一个对之早怀警惧之心的人感到惊慌失措。就连荷马偶尔也会打盹，亚历山大之所以从高位跌落，要归因于他的幻想。事情的成败得失，取决于多种环境——在一种环境下注定胜利的因素，在另外的环境下就会导致失败。在所有事情当中，不可救药的蠢行同样都要归于空无一物的自满。树开花，草开花，藤蔓也开花，全都是为了结籽。

SHOW NO SELF-SATISFACTION.

You must neither be discontented with yourself, which is weak spirited, nor self-satisfied, which is folly. Self-satisfaction arises mostly from ignorance, and it would be a happy ignorance not without its advantages if it did not ruin reputation. Because a person cannot achieve the superlative perfections of others, he contents himself with any mediocre talent of his own. Distrust is wise, and even useful, either to evade mishaps or to afford consolation when they come. For a misfortune cannot surprise a man who has already feared it. Even Homer nods at times, and Alexander fell from his lofty state due to his illusions. Things depend on many circumstances — what constitutes triumph in one set may cause a defeat in another. In the midst of all, incorrigible folly remains the same with empty self-satisfaction, blossoming, flowering, and running all to seed.

www.buhua.com
2004

通向伟大的捷径就是与他人同行

与正直之士交往非常有益；风度举止和格调趣味被互相分享，良好的判断力、甚至还有才能，都在不知不觉地中增长。那么，就让风风火火的人和慢条斯理之辈结成伙伴吧，有其他性情气质的人也不妨如此，这样一来，无需强迫，他的性格就会变得中庸。与他人协调一致是一门大技巧。性质相反的美彼此交替，支撑着这个世界；如果这种方式能促成物质世界的协调，那对精神世界就更是如此。在选择朋友和对手的时候不妨采用这种策略——通过结合两个极端，更有效的中间道路也就找到了。

THE SHORTEST PATH TO GREATNESS IS ALONG WITH OTHERS.

Intercourse with the right people works well; manners and taste are shared, good sense and even talent grow insensibly. Let the impatient person then make a comrade of the sluggish, and so with the other temperaments, so that without forcing it the golden mean is obtained. It is a great art to agree with others. The alternation of contraries beautifies and sustains the world, and if it can cause harmony in the physical world, still more can it do in the moral. Adopt this policy in the choice of friends and defendants — by joining extremes the more effective middle way is found.

不要吹毛求疵

有一些性格阴暗的人,每件事情都能挑出毛病,这倒并不是出于什么邪恶的动机,而是他们的天性使然。他们谴责所有人——那些他们已经为之做过事情的人,那些他们将要为之做事情的人。这显示出一种比残忍更为糟糕的天性，确实很可恶。他们对别人的责难是如此夸张,以至于总是把尘埃说成是足以刺瞎眼睛的光束。他们一直就是能把天堂变成监狱的工头——要是有激情介入,他们就会把事情推向极端。相反,高尚的天性总是懂得如何为失败找到开脱之辞,他们会说,意图是好的,或者说,这是疏忽的错误。

DO NOT BE CENSORIOUS.

There are people of gloomy character who regard everything as faulty, not from any evil motive but because it is their nature to. They condemn all — these for what they have done, those for what they will do. This indicates a nature worse than cruel, vile indeed. They accuse with such exaggeration that they make out of motes beams with which to poke out the eyes. They are always taskmasters who could turn a paradise into a prison — if passion intervenes they drive matters to the extreme. A noble nature, on the contrary, always knows how to find an excuse for failings, saying the intention was good, or it was an error of oversight.

要学会微笑着，歌唱着去
面对生活的苦难哦！
家
工
作

不要等到自己成为落日

智者的座右铭是：在事物抛弃你之前先把它们抛弃掉。一个人应该能在最后的时刻尽力把胜利抓在自己手里，就像太阳常常在它还很明亮的时候就退到云层的后面，这样人们就看不到落日西沉，从而满腹狐疑：它究竟沉下去了没有？当剩下的机会只是灾祸的时候，要明智地全身而退，免得等它变成现实的时候你被迫这样做。不要等到人们对你冷眼相待并把你送进坟墓，感觉依然活着，敬意却已死去。聪明的驯马师会及时把赛马放归牧场，而不会等到它在比赛途中轰然倒下、惹人嘲笑。美人应该老早砸碎自己的镜子，不要等到眼睁睁地看着自己红颜老去，揽镜自照，悔之晚矣。

DO NOT WAIT TILL YOU ARE A SETTING SUN.

It is a maxim of the wise to leave things before things leave them. One should be able to snatch a triumph at the end, just as the sun even at its brightest often retires behind a cloud so as not to be seen sinking, and to leave in doubt whether he has sunk or not. Wisely withdraw from the mere chance of mishap, lest you have to do so when it becomes reality. Do not wait until they turn you the cold shoulder and carry you to the grave, alive in feeling but dead in esteem. Wise trainers put racehorses out to pasture before they arouse derision by falling on the course. A beauty should break her mirror early, lest she do so later with open eyes.

Hi ~

要有朋友

一位朋友就是另一个自己。每个人在他的朋友看来都是善良而睿智的；在他们之间，万事都会变得顺遂。人人都会成为别人希望他成为的那种人——为了让别人希望你好，你就必须赢得他们的好心，赢得他们的嘉言。没有什么魔法比得上友善的言行，而获得友善之情的方法，就是行友善之事。我们当中到大多数最优秀的人，都要依靠他人——我们要么生活在朋友中，要么生活在敌人中。因此，每天都要寻找希望你好的人——即使眼下不是朋友，但在经过考验之后，其中必定有人不久就会成为你的心腹知己。

HAVE FRIENDS.

A friends is a second self. Every friend is good and wise for his friend; between them everything turns to good. Everyone is as others wish him to be — but in order that they may wish him well, he must win their hearts and so their tongues. There is no magic like a good turn, and the way to gain friendly feelings is to do friendly acts. The most and best of us depend on others — we have to live either among friends or among enemies. So seek someone everyday who will wish you well — if not a friend, by and by after trials some of these will become your confidants.

赢得善意

因为这是预见并促进最伟大目标的第一动因和最高动因。通过赢得人们的善意,你就获得了他们的好评。有的人对自己的长处过于信任,以至于忽视了他人的恩惠,但智者懂得:如果没有来自他人的厚爱,人生将是一条漫长而崎岖的道路。善意使每件事情变得更容易,对每件事情都有所助益,这些事情有:勇气,热情,知识,甚或判断力;同时它不会看到缺点,因为它从不找缺点。它来自于某种共同的利益,要么是物质上的,比如志趣、民族、家庭、种族、职业等;要么是形式上的(这是一种更高层次的共有),比如身份、义务、名声和优点。全部的困难就在于赢得善意——保持它并不难。然而,你要去寻求善意,找到之后要善加利用。

GAIN GOODWILL.

For thus the first and highest cause foresees and furthers the greatest objects. By gaining their goodwill you gain people's good opinion. Some trust so much to merit that they neglect grace, but wise men know that it is a long and stony road without a lift from favor. Goodwill facilitates and supplies everything. It supposes gifts or even supplies them, such as courage, zeal, knowledge, or even discretion; whereas it will not see defects because it does not search for them. It arises from some common interest, either material, as in disposition, nationality, family, race, occupation; or formal, which is of a higher kind of communion, as in capacity, obligation, reputation or merit. The whole difficulty is to gain goodwill — to keep it is easy. It has, however, to be sought for and when found to be utilized.

好朋友！
好兄弟！

在兴盛时期要为灾祸作准备

夏天里收集过冬的储备，既更明智，也更容易。春风得意的时候，青睐很廉价，朋友也很多。因此，最好是把它们储存起来留到更不幸的日子，因为，倒霉背运的时候，花费很昂贵，帮手却没有。把朋友和那些欠你情的人储存起来吧——他们价格看涨的日子没准会到来。卑鄙小人从来就没有朋友——得意之时不认人，倒霉之日人不认。

IN TIMES OF PROSPERITY PREPARE FOR ADVERSITY.

It is both wiser and easier to collect winter stores in summer. In prosperity favors are cheap and friends are many. It is well therefore to save them for more unlucky days, for adversity costs dear and has no helpers. Retain a store of friends and people who are in your debt — the day may come when their price will go up. Lowly minds never have friends — in luck they will not recognize them, in misfortune they will not be recognized by them.

白白穷鬼！

毋争

每一场竞争都会损害你的名声。竞争对手为了比你更亮而总是抓住机会让你更暗。很少有战争是可敬的。竞争总是揭露出原本可以用谦恭来掩藏的缺点。许多人在没有竞争对手的时候一直生活在很好的名声中。争斗的热度让已死的丑闻复活，并赋予它新的生命，挖出长埋地底的残骸。竞争从贬低对手开始，到任何能够求助（而不仅仅是应该求助）的地方寻求帮助。当辱骂的武器对他们的目的没起作用的时候（正如经常或大多数时候所发生那样），对手就会寻求报复，至少要用它们来掸掉任何让你丢脸的事情上的灰尘，而这些事情早已被人遗忘。善意之人总是心平气和，那些拥有好名声的高尚之士，都是善意之人。

NEVER COMPETE.

Every competition damages your reputation. Our rivals seize occasion to obscure us so as to outshine us. Few wage honorable war. Rivalry discloses faults that courtesy would hide. Many have lived in good repute while they had no rivals. The heat of conflict revives and gives new life to dead scandals, digging up long-buried skeletons. Competition begins with belittling, and seeks aid anywhere it can, not only where it should. And when the weapons of abuse do not effect their purpose, as often or mostly happens, our opponents seek revenge and use them at least for beating away the dust of oblivion from anything that is our discredit. People of goodwill are always at peace, and those of good reputation and dignity are of goodwill.

你未傻！
你阳痿！

习惯身边人的缺点

习惯身边人的缺点，就好像习惯一张不太好看的面孔。假如这些人依靠你，或者你依靠他们，这样做就更是绝对必要。有些鄙陋之人，我们或者无法忍受，或者离他不得。所以，聪明的家伙会习惯他们，就像习惯丑陋的面孔，这样一来，既然不得不如此，他们的所作所为也就不一定会让你太觉唐突。最初，他们令人生厌；逐渐地，这样的感觉慢慢消失；最后，通过对厌恶感的反躬自省，我们也就容忍了它。

GET USED TO THE FAILINGS OF THOSE AROUND YOU.

Just as you would to an ugly face. It is indispensable if they depend on you, or you on them. There are wretched characters one cannot live with or without. Therefore clever folk get used to them, as to ugly faces, so that they are not obliged to do so suddenly under the pressure of necessity. At first they arouse disgust, but gradually they lose this influence, and reflection provides for disgust puts up with it.

www.buhua.com
2004

只跟值得尊敬的人打交道

你可以信任他们,他们也会信任你。他们的荣誉是他们的行为的最好担保,哪怕是在误会中,因为他们总是根据自己的品格行事。因此,与其战胜无耻之徒,不如与值得尊敬的人争论。你不能跟名誉破产的人打交道,因为他没有什么东西可以为正直做抵押。与他们打交道,没有真正的友谊可言,他们的协议也没有约束力,不管条款看上去多么严格,因为他们没有荣誉感。不要跟这样的人有任何牵扯,因为,如果荣誉不能约束他们的话,那么美德也做不到,因为荣誉是正直的王座。

ONLY ACT WITH HONORABLE PEOPLE.

You can trust them and they you. Their honor is the best surety of their behavior even in misunderstandings, for they always act according to their character. Hence it is better to have a dispute with honorable people than to have a victory over dishonorable ones. You cannot deal well with the ruined, for they have no hostages for rectitude. With them there is no true friendship, and their agreements are not binding, however stringent they may appear, because they have no feeling of honor. Never have anything to do with such people, for if honor does not restrain them, virtue will not, since honor is the throne of rectitude.

不要谈论自己

如果谈论自己，你要么自夸（这是虚荣），要么自责（这是怯懦）。言者自取其辱，闻者徒增不快。如果说在平常交谈中你应当避免这样，那么在正式场合则更是如此，尤其是在公开演说的时候，在那种场合，言行稍有不慎，都是十足的愚蠢。当面谈论某人，同样需要机智圆巧的谎言，避免走向两个极端：阿谀奉承和横加责难。

NEVER TALK ABOUT YOURSELF.

To do so you must either praise yourself, which is vain, or blame yourself, which is weak minded — it is unseemly for the speaker and unpleasant for the listener. And if you should avoid this in ordinary conversation, how much more so in official matters, and above all in public speaking, where every mere appearance of unwisdom really is unwise. The same want of tact lies in speaking of someone in his presence, owing to the danger of going to one of two extremes: flattery or censure.

www.buhua.com
2004

获得谦恭的美名

这足以让你招人喜爱。优雅有礼乃是文明的主要因素——这是一种必定会赢得众人尊敬的神奇魅力，就像粗鲁无礼必定会招致他们的冷眼和反感一样。说到粗鲁无礼，如果它来自傲慢自负，则令人憎恶；若来自糟糕的教养，则叫人鄙视。谦恭一事，宁多勿少；若有所偏倚，则失之公允。谦恭施于对手之间，尤为胆识之明证。其代价殊小，而获益良多——敬人者，人敬之。礼貌和敬意的长处就在于：你把它们呈献给别人，它们就必定属于你。

ACQUIRE THE REPUTATION FOR COURTESY.

This is enough to make you liked. Politeness is the main ingredient of culture — a kind of witchery that wins the regard of all as surely as discourtesy gains their disfavor and opposition. If this latter springs from pride it is abominable, if from bad breeding it is despicable. Better too much courtesy than too little, provided it is not indiscriminate, which degenerates into injustice. Between opponents it is of special worth as a proof of valor. It costs little and helps much — everyone is honored who gives honor. Politeness and honor have this advantage, that they remain with him who displays them to others.

www.buhua.com
2004

避免招人厌恶

要想招人厌恶，无需什么合适的时机——它不请自来。有许多人不由自主地憎恶别人，既不清楚为何憎恶，也不知道如何憎恶。他们的恶意比我们的好心来得更快。他们的恶意比他们热衷于谋取利益的贪婪更容易损害他人。有的人总是想方设法与每个人搞僵，因为他们要么总是招惹、要么正在经历恼恨的心情。憎恨，就像坏名声一样，一旦生根就很难清除。明智之士受人敬畏，恶意之徒招人憎恶，傲慢之人遭人鄙视，小丑被人辱慢，怪人被人忽视。因此，付出尊敬你就会赢得尊敬，要懂得：要想受人尊敬，必须尊敬他人。

AVOID BECOMING DISLIKED.

There is no right occasion to seek dislike — it comes without seeking soon enough. There are many who hate of their own accord without knowing the why or the how. Their ill will outruns our readiness to please. Their ill nature is more prone to do harm to others than their greed is eager to gain advantage for themselves. Some manage to be on bad terms with everyone because they always either produce or experience a vexation of spirit. Once hate has taken root it is, like bad reputation, difficult to eradicate. Wise people are feared, the malevolent are abhorred, the arrogant are regarded with disdain, buffoons with contempt, eccentrics with neglect. Therefore pay respect that you may be respected, and know that to be esteemed you must show esteem.

看什么看！
讨厌！

生活要讲求实际

就连知识也要符合潮流，对不合时宜的知识，假装无知是明智的。思想和品味与时俱变。在思考方式上不要墨守成规，让你的品味与时俱进。在所有事情上，多数人的品味总是占上风；要想引领时代走向更高目标，你就必须紧跟时代。在身体的装饰上，就像灵魂的装饰一样，要让自己适应现在，即使过去的看上去更好。但这一法则不能应用于善，因为善适用于所有的时代。如今它被忽视了，似乎是过时了。诚实真正，信守诺言，以及拥有这样一些品德的好人，似乎都是来自美好的往昔，他们受到所有人的喜爱，但尽管如此，如果有任何一位幸存至今的话，他们也不合潮流，不会被人仿效。如今，美德被看作更怪异，恶行被视为理所当然，这对我们的时代来说是多么不幸的一件事啊！如果你明智，就按照自己所能去生活，如果你做不到，就按照自己所愿去生活。更多地想想命运给予了你什么吧，而不要总是想它拒绝了什么。

LIVE PRACTICALLY.

Even knowledge has to be in style, and where it is not it is wise to affect ignorance. Thought and taste change with the times. Do not be old fashioned in your ways of thinking and let your taste be modern. In everything the taste of the many carries the day; for the time being one must follow it in hope of leading it to higher things. In the adornment of the body, as of the mind, adapt yourself to the present, even though the past appears better. But this rule does not apply to kindness, for goodness is for all times. It is neglected nowadays and seems out of date. Truthfulness, keeping your word, and so too good people, seem to come from the good old days, yet they are liked for all that, but even so if any exists they are not in fashion and are not imitated. What a misfortune for our age that it regards virtue as a stranger and vice as a matter of course! If you are wise live as you can, if you cannot live as you would. Think more highly of what fate has given you than of what it has denied.

早上好！

不要无事忙

正如有些人对每件事情都要闲谈漫议一番一样，另一些人对每件事情都要大费周章。他们总是说大话，认真地对待每一件事情，把它带入争吵不休，或者让它变得高深莫测。棘手的事情，如果能避开的话，就不应该过于认真地对待。把你本该从肩上放下来的东西放在心头上，这是很荒谬的。很多本该有些分量的事，因为丢下不管而变得无足轻重；而那些微不足道的事，却因为大费周章而变得关乎至重。在一开始，事情可以很容易地得到解决，但到后来就难了。治疗常常带来疾病。把事情丢下不管，是很重要的生活法则。

DO NOT MAKE MUCH ADO ABOUT NOTHING.

As some make gossip out of everything, so others make much ado of everything. They always talk big, take everything in earnest and turn it into a dispute or a secret. Troublesome things must not be taken too seriously if they can be avoided. It is preposterous to take to heart that which you should just throw over your shoulders. Much that would be something has become nothing by being left alone, and what was nothing has become of consequence by being made much of. At the outset things can be easily settled, but not afterwards. Often the remedy causes the disease. It is by no means the least of life's rules to let things alone.

快！快！快！
来病人了！
快抢救！！

言辞和行动中的卓越

凭借这一点，你可以在许多地方获得一个位置，并事先就赢得尊敬。它表现在每件事情上，在言谈上，在外表上，甚至在步态上。对征服人心来说，它是一次伟大的胜利。它并非来自任何愚蠢的自以为是或华而不实，而是在于适当的权威口气，而这，则来自更高才能和真正优点的结合。

DISTINCTION IN SPEECH AND ACTION.

By this you gain a position in many places and win esteem in advance. It shows itself in everything, in talk, in look, even in gait. It is a great victory to conquer people's hearts. It does not arise from any foolish presumption or pompous talk, but in a becoming tone of authority born of superior talent combined with true merit.

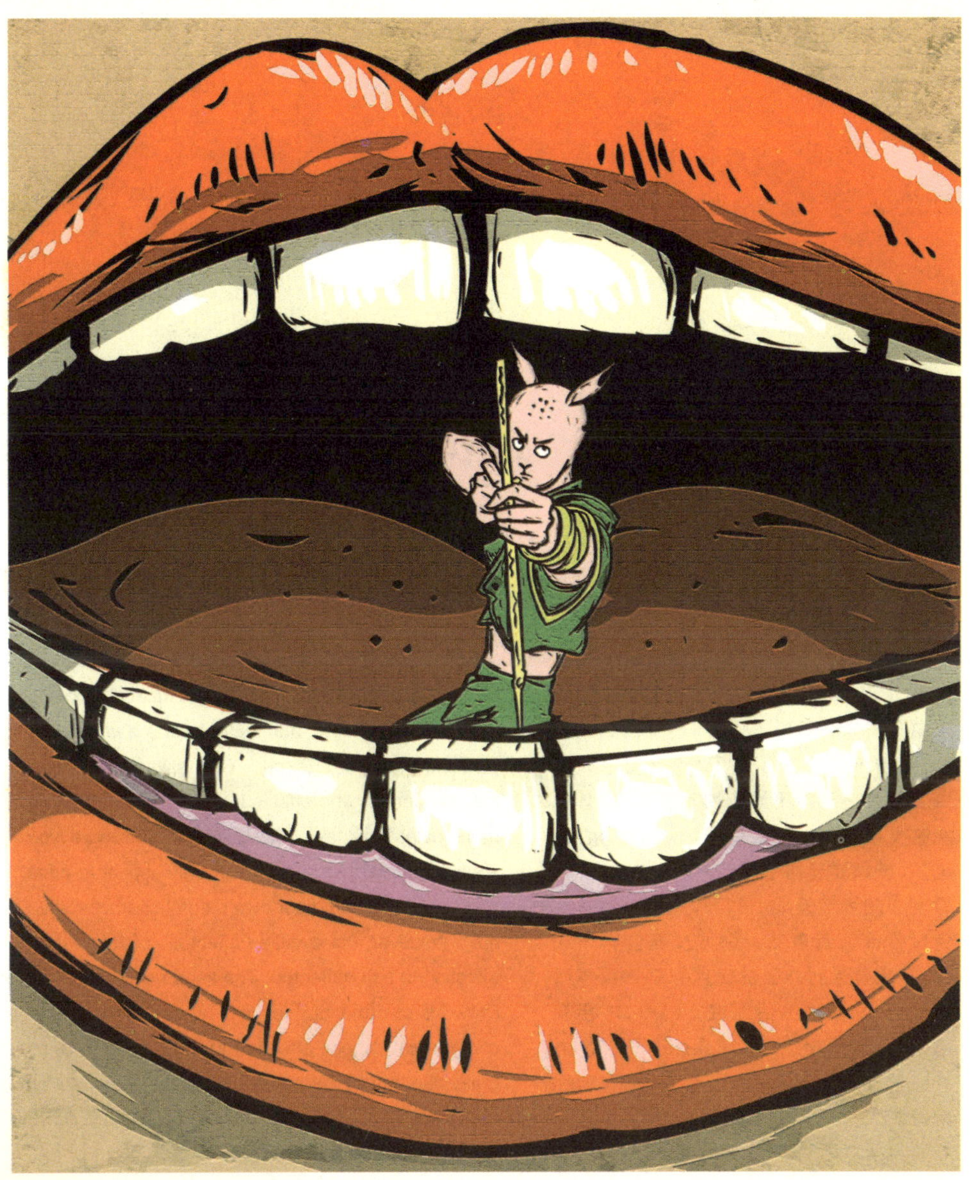

不要装模作样

优点越多，做作越少。对所有人来说，装模作样都显得品味低俗。做作之人，既令他人心生厌恶，也让自己不堪其苦，因为他长期饱受专心料理自己的折磨。就连最突出的优点，也会因为这样的做作而丧失泰半，因为它使得这些优点看上去更像是自鸣得意和矫揉造作，而不是自然的产物，率性自然总是要比矫揉造作更令人愉快。伪装自己并不具备的美德，别人必定能感觉出来。对一件事情，你所费的心力越多，就越要显得举重若轻，这样一来，它看上去就浑然天成。然而，也不要为了避免做作而假装自然，那样反而会弄巧成拙。智者贤人对自己的优点似乎一直懵然不知，只有对这些优点漫不经心，才会唤起旁人对它们的注意。如果所有人都认为你完美无缺，只有你自己不以为然，那你就倍加伟大——你通过两条相反的道路而赢得了欢呼喝彩。

AVOID AFFECTATION.

The more merit, the less affectation, which gives a vulgar flavor to all. It is wearisome to others and troublesome to the one affected, for he becomes a martyr to care and tortures himself with attention. The most eminent merits lost most by it, for they appear proud and artificial instead of being the product of nature, and the natural is always more pleasing than the artificial. One always feels sure that the person who affects a virtue has it not. The more pains you take with a thing, the more you should conceal them, so that it may appear to arise spontaneously from your own natural character. Do not, however, in avoiding affectation fall into it by affecting to be unaffected. The sage never seems to know his own merits, for only by not noticing them can you call others' attention to them. He is twice great who has all the perfections in the opinion of all except of himself — he attains applause by two opposite paths.

www.buhua.com
2004

让自己成为众望所归的人

很少有人从多数人那里获得这样的青睐，如果获得了智者的青睐，那就是最大的快乐。当你完成自己的工作时，淡然处之是一般的法则。但有很多方式可以挣得善意的回报。最稳妥的方式，是在职务和才能上都超群出众；此外就是令人愉快的举止风度，这样你就到达了这样一种状态：对你的职务来说你是必不可少的，而不是职务对你来说必不可少。有的人对他们的职位表示敬意，另一些人则反其道而行之。如果是因为继任者差劲而让前任看上去还算不错的话，那就不是什么了不起的收获了，因为这并不意味着人们想念前任，而是意味着人们希望摆脱后任。

MAKE YOURSELF SOUGHT AFTER.

Few reach such favor with the many, if with the wise it is the height of happiness. When one has finished one's work, coldness is the general rule. But there are ways of earning the reward of goodwill. The sure way is to excel in your office and talents; add to this agreeable manner and you reach the point where you become necessary to your office, not your office to you. Some do honor to their post, with others it is the other way around. It is no great gain if a poor successor makes the predecessor seem good, for this does not imply that the one is missed, but that the other is wished away.

不要做他人过错的记录者

老是惦记别人的恶名，意味着自己也有个不光彩的名声。有些人希望用别人的污点来遮掩自己的污点，或者至少是希望摆脱这些污点；抑或想从这里寻找安慰——这是愚蠢的安慰。那些充当全城的丑闻下水道的人，他们的呼吸必定臭不可闻。这样的事情挖得越多，自己就被弄得越脏。很少人没有污点。只有那些默默无闻的人，其过错才罕为人知。那么，留心避免做一个过错记录者吧。那将是一件令人憎恶的事，一个活得没有心灵的人。

DO NOT BE A BLACKLISTER OF OTHER PEOPLE'S FAULTS.

It is a sign of having a tarnished name to concern oneself with the ill fame of others. Some wish to hide their own stains with those of others, or at least wash them away; or they seek consolation therein — it is the consolation of fools. Their breath must stink who form the sewers of scandal for the whole town. The more one grubs about in such matters the more one befouls oneself. There are few without stain somewhere or other. It is only of little known people that the failings are little known. Be careful then to avoid being a registrar of faults. That is to be an abominable thing, a man that lives without a heart.

愚蠢并不在于干蠢事，而在于干了蠢事不知藏拙

你应该对自己的愿望秘而不宣，更应该让自己的过失不为人知。人非圣贤，孰能无过。智者善藏己拙，愚夫自夸其短。名声的得来，更多地依赖于你的所藏，而不是你的所为；一个人如果不能成为谦谦君子，他一定要谨小慎微。巨匠之失，日月其蚀。即便是亲朋之间，向一个人的朋友暴露他的弱点，也极不寻常。不仅如此，倘若可能，你甚至应该向自己隐瞒自己的弱点。其实这并不难做到，你只要记住另一条人生至理：学会遗忘。

FOLLY CONSISTS NOT IN COMMITTING FOLLY, BUT IN NOT HIDING IT WHEN COMMITTED.

You should keep your desires sealed up, still more your defects. All go wrong sometimes, but the wise try to hide their errors while fools boast of them. Reputation depends more on what is hidden than on what is done; if a man does not live chastely, he must live cautiously. The errors of great men are like the eclipses of the greater lights. Even in friendship it is rare to expose one's failings to one's friend. Nay, one should conceal them from oneself if one can. But here one can help with that other great rule of life: learn to forget.

www.buhua.com
2004

凡事优雅从容

优雅，是天才的生命，言辞的呼吸，行动的灵魂，装饰中的装饰。完美，乃是我们天性的装饰，而优雅，则是完美本身的装饰。甚至在思想中，我们也能找到它的身影。优雅的风度，源自天性者殊多，得之教育者殊少——它甚至胜过后天的训练。它可以更轻松自如地走向自由，更悠闲自在地克服阻碍，给完美添上最后的一笔。如果没有优雅，纵使美奂美伦，亦生气全无；就算毕恭毕敬，也风度尽失。它超过勇敢、判断力、审慎，乃至帝王之尊。它是走向成功的捷径，也是摆脱困境的妙方。

GRACE IN EVERYTHING.

It is the life of talent, the breath of speech, the soul of action, and the ornament of ornament. Perfections are the adornment of our nature, but this is the adornment of perfection itself. It shows itself even in the thoughts. It is mostly a gift of nature and owes least to education — it even triumphs over training. It is more than ease, approaches the free and easy, gets over embarrassment, and adds the finishing touch to perfection. Without it beauty is lifeless, graciousness ungraceful. It surpasses valor, discretion, prudence, even majesty itself. It is a shortcut to accomplishment and an easy escape from embarrassment.

www.buhua.com
2004

品格高尚

这是成为一个绅士的首要条件之一，它鞭策我们走向各种高贵的情怀。它使趣味得以改进，使心灵更加高贵，它提升精神，净化情感，强化尊严。它提升拥有这种品格的人。有时候，幸运之神因为妒嫉而转身离去，它甚至能矫正这种不利的转向。当意志不能在行动中发挥作用的时候，高尚的品格就可以找到施加影响的机会。宽厚、慷慨，以及一切英雄的品质，都把它认作是自己的源泉。

HIGH-MINDEDNESS.

This is one of the principal qualifications for a gentleman, it spurs us on to all kinds of nobility. It improves the taste, ennobles the heart, elevates the mind, refines the feelings, and intensifies dignity. It raises him in whom it is found. At times it even remedies the bad turns of fortune, which turns itself around because of envy. High-mindedness can find full scope in the will when it cannot be exercised in act. Magnanimity, generosity, and all heroic qualities recognize in it their source.

还是高尚
最美！

决不诉苦

诉苦者总是自取其辱。成为一个反抗他人凌辱的自立模范，总比成为他们的同情对象要强得多。诉苦，只会鼓励那些听我们诉苦的人去效法我们所控诉的人，透露一次欺侮，只会为另一次欺侮制造藉口。控诉过去受到的伤害，就给未来的伤害创造了机会。我们寻求别人的支持和忠告，所得到的，只能是冷漠和轻蔑。更高明的策略，是赞美他人对你的恩惠，这样，闻之者或许觉得有仿效的必要。讲述我们从那些不在场的人那儿所得到的恩惠，也就是要求那些在场的人要有同样的付出，这样，我们也就把此人对我们的信任推销给了彼人。是故，明智之士从不会把自己的失败或缺陷公之于众，而只是张扬那些让自己赢得尊重的评价，这样的评价，使亲者快，让仇者痛。

NEVER COMPLAIN.

To complain always brings discredit. Better to be a model of self-reliance opposed to the passion of others than an object of their compassion. For complaining opens the way for the hearer to act like those we are complaining of, and to disclose one insult forms an excuse for another. By complaining of past offenses we give occasion for future ones, and in seeking aid or counsel we only obtain indifference or contempt. It is much more politic to praise a person's favors, so that others may feel obliged to follow suit. To recount the favors we owe the absent is to demand similar ones from those present, and thus we sell our credit with the ones to the other. The shrewd will therefore never publish to the world his failures or his defects, but only those marks of consideration that serve to keep friendship alive and enmity silent.

www.buhua.com
2004

既要苦干，也要让你的苦干为人所知

人们对事物的看法，不据其所是，而依其所见。有用之才，如果知道如何展示自己的有用之处，就会倍加有用。不见之物，犹如乌有。正义之事，如果看上去不正义，就得不到应有的尊重。明察秋毫者盖寡，蒙蔽于表象者殊多。欺骗无处不在——论人衡物，皆凭外表；盛装之下，其实不副。不过，一个好的外表，端的是其内在完美最好的推荐信。

DO AND BE SEEN DOING.

Things do not pass for what they are but for what they seem. To be of use and to know how to show it, is to be twice as useful. What is not seen is as if it was not. Even the right does not receive proper consideration if it does not seem right. The observant are far fewer in number than those who are deceived by appearances. Deceit rules — things are judged by their jackets and many things are other than they seem. But a good exterior is the best recommendation of the inner perfection.

www.buhua.com
2004
大城门 ××造

高尚的情操

有一种灵魂的卓越，高尚的品格激发英勇的行为，赋予整个品格以优雅的气派。高尚的情操并不常有，因为它必须先有宽阔的胸襟。它的主要特点是对敌人有良言，甚至更有良行。当复仇的机会来临的时候，它焕发的光彩最明亮；不是单方面把机会放过，而是借助一次全面胜利来利用这个机会，以展示意想不到的宽宏大度。这是精妙的策略之举——不，是治国方略的极致。它从不炫耀胜利，它装作若无其事，在赢得奖赏的同时，它对自己的功绩藏而不露。

NOBILITY OF FEELING.

There is a certain distinction of the soul, a high-mindedness prompting to gallant acts, that gives an air of grace to the whole character. It is not found often, for it presupposes great magnanimity. Its chief characteristic is to speak well of an enemy and to act even better toward him. It shines brightest when a chance comes for revenge; not alone does it let the occasion pass but improves it by using a complete victory in order to display unexpected generosity. It is a fine stroke of policy — no, the very acme of statecraft. It makes no pretense to victory, for it pretends to nothing, and while obtaining its deserts it conceals its merits.

真正的冠军还
应该是您！

三思而行

向内心的再审法庭上诉，就能让事情变得更稳妥。尤其是当行动路线尚不清晰的时候，你要争取时间来确认或改进你的决定。它为你巩固或确证自己的判断提供了回旋的余地。如果是赠予之事，那些明显经过慎重考虑的礼物，要比仓促出手的礼物更受人重视；长久的期待会得到最高的奖赏。如果不得不拒绝，你要争取时间以决定在什么时候、用什么方式说“不”，以便让它变得更容易接受。此外，在渴望得到的热情消退之后，对拒绝的憎恶感就会显著减少。尤其是当人们迫切要求答复的时候，你最好是推迟答复，因为往往只要虚晃一招就能消除别人的关注。

REVISE YOUR JUDGEMENTS.

To appeal to an inner court of revision makes things safe. Especially when the course of action is not clear, you gain time either to confirm or improve your decision. It affords new grounds for strengthening or corroborating your judgement. And if it is a matter of giving, the gift is the more valued from its being evidently well considered than for being to promptly bestowed; long expected is highest prized. And if you have to deny something, that gains you time to decide how and when to mature the no so that it may be made palatable. Besides, after the first heat of desire is passed the repulse of refusal is felt less keenly. But, especially when people press for a reply, it is best to defer it, for as often as not that is only a feint to disarm attention.

凡事都慢三拍哦！

宁与他人一起疯狂，也不一个人独自清醒

政治家们都这么说。如果人人都疯狂的话，多一个疯狂的人并不会更糟，反之，独善其身的智者反倒被视为愚蠢。重要的是随波逐流。最高的智慧，常常就包括无知，或者假装无知。你不得不与他人一起生活，而他人大部分都是无知的。“要完全离群索居，你必须要么完全像上帝，要么彻底像野兽。”但我要把这句格言改一下：与其一人独傻，不如与大家同智。也有人追逐奇思妙想，为的是要标新立异。

BETTER MAD WITH THE REST OF THE WORLD THAN WISE ALONE.

So say politicians. If all are so, one is no worse off than the rest, whereas solitary wisdom passes for folly. So important is it to sail with the stream. The greatest wisdom often consists of ignorance, or the pretense of it. One has to live with others, and others are mostly ignorant. “To live entirely alone one must be very like a god or quite like a wild beast,” But I would turn the aphorism by saying: Better be wise with the many than a fool all alone. There be some too who seek to be original by chasing chimeras.

储备双倍的资源

这样你就可以使人生倍加丰富。一个人不必依赖于一件事情，也不要仅仅依靠一种才智，无论这一才智如何突出。每样东西都应该保存双份，尤其是关乎到成功、恩惠和尊敬等方面的缘由。月有阴晴圆缺，事有成败利钝，天下万有，莫此为甚。不过最变幻莫测的，还是那些依赖于人类意志的事情，而人类的意志，乃是所有事情中最脆弱的东西。要防备这样的变化无常，就要有智者之虑，为此，就要保存双份的优良有益的品质储备，这是生活的首要法则。正如大自然对我们最重要的手足以及那些最冒风险的器官都给了双份一样，我们也要以这样技巧对待那些我们赖以成功的品质。

DOUBLE YOUR RESOURCES.

You thereby double your life. One must not depend on one thing or trust to only one resource, however preeminent. Everything should be kept double, especially the causes of success, of favor, or of esteem. The moon's mutability transcends everything and gives a limit to all existence, especially of things dependent on human will — the most brittle of all things. To guard against this inconstancy should be the sage's care, and for this the chief rule of life is to keep a double store of good and useful qualities. Thus as nature gives us in duplicate the most important of our limbs and those most exposed to risk, so art should deal with the qualities on which we depend for success.

www.buhua.com
2004

不要唱反调

老唱反调，只能证明你愚蠢而乖戾，审慎之人应该提防这样的紧张。凡事都挑毛病或许能证明你聪明，但此类口舌之争会使你被人们视为一个傻瓜。这样的人，让最愉快的交谈成为一场战争，这样一来，对于他们的同事朋友，较之于那些和他们毫无瓜葛的人，他们更像是死对头。美味佳肴中碰上砂石把牙齿磨得嘎吱作响，最叫人不能忍受；在轻松愉快的时候抬杠，也正是如此。用轭把野兽和家畜套在一起，这种人既愚蠢又残忍。

DO NOT NOURISH THE SPIRIT OF CONTRADICTION.

It only proves you foolish or peevish and prudence should guard against this strenuously. To find difficulties in everything may prove you clever but such wrangling writes you down as a fool. Such folk make a war out of the most pleasant conversation and in this way act as enemies toward their associates rather than toward those with whom they do not consort. Grit grates most in delicacies, and so does contradiction in amusement. They are both foolish and cruel who yoke together the wild beast and the tame.

www.buhua.com
2004

把自己置于事情的核心

这样你就可以感受到事务的脉动。很多人误入歧途，迷失于无益讨论的岔道和喋喋不休的灌木丛，永远也认识不到手头事情的真正本质。他们总是纠缠于一点，虽百遍而不厌，让自己和他人疲于奔命，却从未触及事务的重要核心。这源自于他们自己无法挣脱的头脑混乱。他们总是把时间和耐性浪费在那些本该丢下不管的事情上，然后却没有多余的时间去做他们已经丢下不管的事。

POST YOURSELF IN THE CENTER OF THINGS.

So you feel the pulse of affairs. Many lose their way either in the ramifications of useless discussion or in the brushwood of wearisome verbosity without ever realizing the real matter at hand. They go over a single point a hundred times, wearing themselves and others, and yet never touch the all important center of affairs. This comes from a confusion of mind from which they cannot extricate themselves. They waste time and patience on matters they should leave alone, and afterward there is no time spared for what they have left alone.

水还越来越多
了啊？！

贤者自足

集万有于己身者，一切都随身携带。一位全知全能的朋友，能够对罗马和世界的其余部分代表我们，让你成为自己的这样一位朋友吧，那么你就能够独自生活了。这样一个人会需要谁呢？如果他并不比自己更聪明、更有品位的话。那么，你就会只依靠自己，这是最高的快乐，就像是至高的上帝一样。能够独立生活的人，与野兽毫无相似之处，与圣哲贤士相似之处颇多，在每件事情上都像神一样。

THE SAGE SHOULD BE SELF-SUFFICIENT.

He that was all in all to himself carried all with him when he carried himself. If a universal friend can represent us to Rome and the rest of the world, let a man be his own universal friend, and then he is in a position to live alone. Whom could such a man want if there is no clearer intellect or finer taste than his own? He would then depend on himself alone, which is the highest happiness and like the Supreme Being. He that can live alone resembles the brute beast in nothing, the sage in much and like a god in everything.

那也只有我了！

顺其自然

尤其当公众或私人生活掀起狂波骤浪的时候，更是如此。有世事的狂风，有激情的暴雨，此时，明智的做法是：撤退至宁静的港湾，停泊在那里等待风平雨歇。治疗常使疾病恶化，在这种情况下，你只有把疾病交给自然的疗程和时间的道德感化。明智的大夫，知道何时不用开药，有时候，不下一药，更称妙手。平息俗世风暴的正确方法，就是袖手旁观，任其自生自灭。眼下退避三舍，稍后战而胜之。一汪清泉，稍稍搅动就会浑浊，要想让它清澈如故，你不能动手动脚，而只能不闻不问。顺其自然，听其自灭，治乱之法，于斯为善。

THE ART OF LETTING THINGS ALONE.

The more so the wilder the waves of public or of private life. There are hurricanes in human affairs, tempests of passion, when it is wise to retire to a harbor and ride it out at anchor. Remedies often make diseases worse; in such cases one has to leave them to their natural course and the moral influence of time. It takes a wise doctor to know when not to prescribe, and at times the greater skill consists in not applying remedies. The proper way to still the storms of the vulgar is to hold yourself back and let them calm down by themselves. To give way now is to conquer by and by. A fountain gets muddy with but little stirring up, and does not get clear by our meddling with it but by our leaving it alone. The best remedy for disturbances is to let them run their course, for so they quiet down.

www.buhua.com
2004

正视倒霉的日子

人人都有倒霉的时候，这是一种客观存在。这种时候诸事不顺，即使再换一局，手气仍然很臭。一件事试上两次，就足以知道今天是幸运还是倒霉。万物皆在变化之中，就连心智也是如此，没有人永远聪明。机遇至关重要，即使是如何写一封好信这样的事情也不例外。所有的完美都适时而至，就连美景也并非四季常在。智者虽有千虑，有时也马失前蹄，或由太过，或因不及。要想事情完美，也须适得其时。为什么有些人事事皆不如意，而有些人样样都很顺利，其道理亦正在此。你发现一切准备就绪，你才思敏捷，你天资卓异，你的幸运之星正在冉冉升起。这种时候，你必须抓住机会，一刻也不要浪费。不过，明智之士不会仅凭一件事情的好坏来判定这一天是幸运还是倒霉，因为一件好事可能仅仅是机缘凑巧，而一件坏事也不过是一桩小小的烦恼。

RECOGNIZE UNLUCKY DAYS.

They do exist. Nothing goes well on them, and even though the game may be changed the bad luck remains. Two tries should be enough to tell if one is in luck today or not. Everything is in process of change, even the mind, and no one is always wise. Chance has something to say, even how to write a good letter. All perfection turns on the times — even beauty has it hours. Even wisdom fails at times by doing too much or too little. To turn out well a thing must be done on its own day. This is why with some people everything turns out ill, with others all goes well, even with less trouble. They find everything ready, their wit prompt, their presiding genius favorable, their lucky star on the rise. At such times one must seize the occasion and not throw away the slightest chance. But a shrewd person will not decide on a day's luck by a single piece of good or bad fortune, for the one may be only a lucky chance and the other a slight annoyance.

www.buhua.com
2004

立刻发现事情中好的一面

这是良好品味的优势。蜜蜂为她的蜂房而酿蜜,毒蛇为它的毒液而酿毒。品味也是如此——有人追香,有人逐臭。任何事情中都有好的东西,尤其是书籍,因为它为思想提供食粮。许多人有这样一种嗅觉:在一千个优点里他们总是盯着一个缺点,把它挑出来就是为了责备,仿佛他们就是人的心灵和精神的清道夫。他们就是这样编订缺点的资产负债表,这张负债表,把更多的信用委托给他们糟糕的品味而不是他们的智力。他们过着悲哀的生活,用苦涩滋养自己,用垃圾肥沃自己。那些在一千个缺点中抓住他们碰巧发现的一点美的人,有着更幸运的品味。

FIND THE GOOD IN A THING AT ONCE.

This is the advantage of good taste. The bee goes to the honey for her comb, the serpent to the gall for its venom. So with taste — some seek the good, others the ill. There is nothing that has no good in it, especially in books, as giving food for thought. But many have such a scent that amid a thousand excellences they fix upon a single defect, and single it out for blame as if they were scavengers of people's hearts and minds. So they draw up a balance sheet of defects, which does more credit to their bad taste than to their intelligence. They lead a sad life, nourishing themselves on bitters and fattening on garbage. They have the luckier taste who amid a thousand defects seize upon a single beauty they may have hit upon by chance.

全年考试成績
50 51 49 52 50 46 42 49
不及格 不及格 不及格 不及格 不及格 不及格 不及格 不及格
至少成绩还挺稳定的！
稳

别自说自听

如果不能取悦他人的话，取悦自己毫无用处。通常，遭人鄙视是对自满的惩罚。一心关注自己，多半会亏欠他人。自言自语、自说自听，不可能产生好的结果。如果说独处的时候跟自己说话是疯狂的话，那么当着他人的面自说自听就是双倍的愚蠢了。交谈的时候老是重复“正像我所说的”和“怎么样？”这是大人物的缺点，会把听众给弄糊涂。他们每说一句话，都要寻求喝彩或捧场，让智者的耐心不堪重负。自负浮夸之徒也是这样跟回声说话，谈话的时候，他们只有借助高跷才能踉踉跄跄地继续下去——每句话都需要一声愚蠢的“好啊！”来支撑。

DO NOT LISTEN TO YOURSELF.

It is no use pleasing yourself if you do not please others, and as a rule general contempt is the punishment for self-satisfaction. The attention you pay to yourself you probably owe to others. To speak and at the same time to listen to yourself cannot turn out well. If to talk to oneself when alone is madness, it must be doubly unwise to listen to oneself in the presence of others. It is a weakness of the great to talk with a recurrent "As I was saying" and "What?" which bewilders their hearers. At every sentence they look for applause or flattery, taxing the patience of the wise. So too the pompous speak with an echo, and as their talk can only totter on with the aid of stilts — at every word they need the support of a stupid "Bravo!"

不要因为对手占了上风你就固守下风

这样的话，你刚一开始就未战先败，并很快就会卸甲丢盔、落荒而逃。抢得上风在对方来说自是精明之举，但你跟在后面固守下风就是愚蠢的了。这样的固执，表现在行动上比表现在言词上更危险，因为做比说冒的风险更大。正是固执的这种共同缺点，使得他们因抵触而失去真实的东西，因争执而失去有用的东西。圣哲贤士从不站在激情的一边，而是支持正确的事业，无论是最初的发现，还是后来的改进。如果敌人是个笨蛋，在这样的情况下他就会转向反面，遵循错误的路线。因此，要想把他赶出上风的位置，就得自己去占据上风，因为他的愚蠢导致他主动放弃上风，而他的固执将让他因此而受到惩罚。

NEVER FROM OBSTINACY TAKE THE WRONG SIDE BECAUSE YOUR OPPONENT HAS ANTICIPATED YOU BY TAKING THE RIGHT ONE.

You begin the fight already beaten and must soon take to flight in disgrace. With bad weapons one can never win. It was astute in the opponent to seize the better side first, it would be folly to come lagging after with the worst. Such obstinacy, is more dangerous in actions than in words, for action encounters more risk than talk. It is the common failing of the obstinate that they lose the true by contradicting it, and the useful by quarrelling with it. The sage never places himself on the side of passion, but espouses the cause of right, either discovering it first or improving it later. If the enemy is a fool, he will in such case turn round to follow the opposite and worse way. Thus the only way to drive him from the better course is to take it yourself, for his folly will cause him to desert it, and his obstinacy be punished for so doing.

不要为了避免陈腐而变得悖谬

NEVER BECOME PARADOXICAL IN ORDER TO AVOID BEING TRITE.

陈腐和悖谬都会极大地损害我们的名声。每一项违背常理的事业都接近于愚蠢。悖谬就是欺骗；它起初因为新奇和刺激而博得喝彩，但到后来，当欺骗被看穿、空虚开始显现的时候，它就变得名誉扫地。它是一种戏法，在政治事务中，将是国家的覆亡。那些不能或不敢经由美德的直路走向伟大行为的人，就会经由悖谬之路绕圈子，受到蠢人的赞佩，却让聪明人成为真正的先知。悖谬表明了不可靠的判断力，即使不完全是基于错误的判断，也肯定是建立在靠不住的基础上，要冒使生活事务更繁重的危险。

Both extremes damage our reputation. Every undertaking that differs from the reasonable approaches foolishness. The paradox is a cheat; it wins applause at first by its novelty and piquancy, but afterwards it becomes discredited when the deceit is foreseen and its emptiness becomes apparent. It is a species of jugglery, and in political matters it would be the ruin of the state. Those who cannot or dare not reach great deeds on the direct road of excellence go round by way of paradox, admired by fools but making wise men true prophets. It demonstrates an unbalanced judgement, and if it is not altogether based on the false, it is certainly founded on the uncertain, and risks the weightier matters of life.

大陆第一老年
内衣模特！

从别人的目标开始，以自己的目标结束

这是一种达到你自己目标的策略性手段。即使在神圣的事情上，基督徒的导师们也总是将重点放在这样神圣的诡计上。这是一种重要的掩饰，这一预见性优势可以用来影响他人的意志。看上去似乎是他的事情已经准备就绪，事实上却是在为你自己的事情开路。除非有了掩护，否则不要冒险前行，尤其当前方危机四伏的时候，更是如此。同样，对付那些总是开口就说“不”的人，最有效的办法就是：提出你自己的目标，免得他们说“不”，以这样一种方式，作出让步的困难也就不会出现在他们身上了。这一忠告属于那种深思熟虑的生活法则，它可以掩饰那些最狡猾的伎俩。

BEGIN WITH ANOTHER'S TO END WITH YOUR OWN.

This is a politic means to your own end. Even in heavenly matters Christian teachers lay stress on this holy cunning. It is a weighty piece of dissimulation, for the foreseen advantages serve as a lure to influence the other's will. His affair seems to be in train when it is really only leading the way for your own. One should never advance unless under cover, especially where the ground is dangerous. Likewise with persons who always say no at first, it is useful to ward off this blow by presenting your intent in such a way that the difficulty of conceding does not occur to them. This advice belongs to the rule about second thoughts, which covers the most subtle maneuvers of life.

www.buhua.com
2004

藏起你受伤的手指，否则凡事都会碰壁

不要就此事向别人诉苦，因为恶念总是瞄准那些容易受伤的软弱之处。愁眉苦脸于事无补，成为别人议论的笑柄只会让你更加愁眉苦脸。恶意总在搜寻伤口以便加以刺激，标枪总是试图瞄准那易怒的脾气，千方百计要刺痛那伤口处的嫩肉。明智之士从不坦白承认所遭受的打击，也不会透露任何不幸，不管是个人的，还是祖上的。就连命运之神，有时候也喜欢伤害我们身上那些最软弱的部位，而且总是伤害已经受伤的肌肉。因此，绝不能暴露你的痛苦之根，也不要宣示你的快乐之源，如果你希望前者很快终止而后者持续绵长的话。

DO NOT SHOW YOUR WOUNDED FINGER, FOR EVERYTHING WILL KNOCK UP AGAINST IT.

Do not complain about it, for malice always aims where weakness can be injured. It is no use to be vexed; being the butt of the talk will only vex you the more. Ill will searches for wounds to irritate, aims darts to try the temper, and tries a thousand ways to sting to the quick. The wise never confess to being hit, or disclose any evil, whether personal or hereditary. For even fate sometimes likes to wound us where we are most tender. It always mortifies wounded flesh. Never therefore disclose the source of pain or of joy, if you wish the one to cease and the other to endure.

www.buhua.com
2004

看透事物的本质

LOOK INTO THE INTERIOR OF THINGS.

事物通常不同于它们看上去的样子。从未看到外壳之下的那种无知，当你展示出内核的时候就会顿然醒悟。谎言总是捷足先登，凭借它们不可救药的粗俗拖着蠢人前行。真相总是姗姗来迟，挽着时间的手臂蹒跚前行。因此，智者总是把他们的两只耳朵中的一只保留给真相，幸好它们共同的母亲——大自然——明智地给出双份。欺骗是非常浅薄的，因此浅薄之徒很容易陷身其中。审慎的生活总是退隐在幽深之处的内部，只有圣哲贤士才能叩访。

Things are generally other than they seem, and ignorance that never looks beneath the rind is disillusioned when you show the kernel. Lies always come first, dragging fools along by their irreparable vulgarity. Truth always lags last, limping along on the arm of time. The wise therefore reserve for truth one of their ears, which their common mother, nature, has wisely given in duplicate. Deceit is very superficial, and the superficial therefore easily fall into it. Prudence lives retired within its recesses, visited only by sages and wise men.

ZZZZZ

不要难于接近

没人会十全十美，以至于不需要旁人不时的忠告。从来都听不进任何意见的人，实在是一个不可救药的笨蛋。即便是智冠群伦之辈，也能为友善的忠告留下方寸之地。至高无上的君王，也须学会屈尊。有些人之所以不可救药，只不过是因为他们不可接近。他们之所以陷于崩溃，是因为没人敢施以援手。即便是至高无上者，也应该为友谊留下敞开的门扉；事实或许会证明：那正是帮助之门。一个朋友，必会毫无保留地给出忠告，甚至是责备，并且不会因此感到窘迫尴尬。我们对他深感满意，我们对他信赖有加，这些坚定的信念给了他这样的权力。一个人大可不必把尊敬和信任随意施予任何人，但是，在他审慎的内心深处，必须有一位知己的真实反照，藉着这一反照，我们能改正自己的错误，并对此心存感激。

DO NOT BE INACCESSIBLE.

None is so perfect that he does not need at times the advice of others. He is an incorrigible ass who will never listen to anyone. Even the most surpassing intellect should find a place for friendly counsel. Sovereignty itself must learn to lean. There are some that are incorrigible simply because they are inaccessible. They fall to ruin because none dares to extricate them. The highest should have the door open for friendship; it may prove the gate of help. A friend must be free to advise, and even to upbraid, without feeling embarrassed. Our satisfaction in him and our trust in his steadfast faith give him that power. One need not pay respect or give credit to everyone, but in the innermost sanctum of his caution a person must have the true mirror of a confidant to whom he owes the correction of his errors, and has to thank for it.

www.buhua.com
2004

要有谈话的艺术

那是展示真实个性的地方。没有比这更需要注意的行为了，因为它是生活中最普通的事情。得失成败，端赖于此。一封书信只不过是一篇深思熟虑的书面谈话而已，如果写封信也要慎重对待的话，那么，有机会即时展现聪明才智的平常谈话，又该要如何加倍小心呢？行家里手能够在言词中感受灵魂的脉动，圣人云："听其言，知其人。"道理也正在此。有人坚持认为，谈话的艺术就是没有艺术——它应该简洁，而不是华而不实，就像衣服一样。这种方式很适合朋友之间的交谈。但在跟那些你应该表示尊重的人交谈的时候，谈话的方式就应该更有尊严，以符合对方的尊贵。要想言词得体，就要适应他人的精神气质和语调口吻。不要咬文嚼字，否则你就会被视为爱炫耀学问的迂夫子；也不要充当观点的收税人，否则人们就会躲着你，或者至少也会把他们的思想卖得很贵。在交谈中，判断力比口才更重要。

HAVE THE ART OF CONVERSATION.

That is where the real personality shows itself. No act requires more attention, thought it be the most common thing in life. You must either lose or gain by it. If it takes care to write a letter, which is but a deliberate and written conversation, how much more so the ordinary kind in which there is occasion for a prompt display of intelligence? Experts feel the pulse of the soul in the tongue, which is why the sage said, "Speak, that I may know thee." Some hold that the art of conversation is to be without art — that it should be neat, not gaudy, like clothing. This holds good for talk between friends. But when held with persons to whom one would show respect, it should be more dignified to answer to the dignity of the person addressed. To be appropriate it should adapt itself to the mind and tone of others. And do not be a critic of words, or you will be taken for a pedant; nor a tax-gatherer of ideas, or people will avoid you, or at least sell their thoughts dear. In conversation discretion is more important than eloquence.

懂得如何诿过于人

对于一个统治者来说，抵挡恶意是一项非凡的技能。让他人接受不满的责难和普遍厌恶的惩罚，并不是一种无能的手段（像某些幸灾乐祸者所想象的），而是一种更为高级的策略。并不是每件事情都尽如人意，因此，即使要付出自尊的代价，也要找一个这样的替罪羊，一个为不幸的事业而承受攻击的靶子。

KNOW HOW TO PUT OFF ILLS ON OTHERS.

To have a shield against ill will is a great piece of skill in a ruler. It is not the resort of incapacity, as ill-wishers imagine, but is due to the higher policy of having someone to receive the censure of the disaffected and the punishment of universal dislike. Everything cannot turn out well, therefore, even at the cost of our pride, to have such a scapegoat, a target for unlucky undertakings.

懂得如何要价

任何物品，仅有内在价值是不够的，因为并非人人都能吃透本质或看穿内部。多数人随大流，别人去哪儿他去哪儿。展示物品的真正价值是一门大技巧——有时候要借助赞美（因为赞美唤起欲求），有时候要靠给它们一个引人注目的名字（这一招对于抬高价格很管用），但不可矫揉造作。另外，声称只卖给行家，通常也是一种诱导方式，因为人人都认为自己是行家，即便不是这样，稀缺感也会唤醒需求。千万别把物品称呼得稀松平常——那会让它们贬值，而不是让它们更容易被人接受。人人都追逐不同寻常的东西，无论对于品味，还是对于智力，罕见之物都更开胃。

KNOW HOW TO GET YOUR PRICE FOR THINGS.

Their intrinsic value is not sufficient, for not everyone bites at the essence or looks into the interior. Most go with the crowd, and go because they see others go. It is a great stroke of art to show things at true value — at times by praising them (for praise arouses desire), at times by giving them a striking name (which is very useful for putting things at a premium), provided it is done without affectation. Again, it is generally an inducement to profess to supply only to connoisseurs, for all think themselves such, and if not, the sense of want arouses the desire. Never call things easy or common — that makes them depreciated rather than made accessible. All rush after the unusual, which is more appetizing both for the taste and for the intelligence.

全球首位
368栖全能
艺人是也。

未雨绸缪

今天是为明天作准备，甚至是为许多天之后作准备。最高明的远见就在于未雨绸缪。深谋远虑者没有无妄之灾,小心谨慎者无需狭路逃生。我们一定不要拖到火烧眉毛的时候才开始思考。深思熟虑可以克服最令人生畏的困难。枕头是沉默的女先知,与其事后辗转反侧,不如事前睡个踏实。许多人先行而后思——换句话说，他们想得更多的是借口,而非后果。有的人则事前事后都不考量。人的一生应该是一个思考的过程，要想想怎样才不至于错过正途。事后的反思和事前的预见,让你能够决定生活的方向。

THINK BEFOREHAND.

Today for tomorrow, and even for many days hence. The greatest foresight consists in determining beforehand the time of trouble. For the provident there are no mischances and for the careful no narrow escapes. We must not put off thought till we are up to the chin in mire. Mature reflection can get over the most formidable difficulty. The pillow is a silent Sibyl, and it is better to sleep on things beforehand than lie awake about them afterwards. Many act first and then think later — that is, they think less of consequences than of excuses. Others think neither before nor after. The whole of life should be one course of thought how not to miss the right path. Rumination and foresight enable one to determine the course of life.

财商
理财

不要与比你优秀的人为伴

NEVER HAVE A COMPANION WHO OUTSHINES YOU.

一个人越优秀，就越不是个令人惬意的伙伴。他愈是出类拔萃，就愈发声名卓著；他总会演奏首席提琴，而你却老是叨陪末座。即使你得到任何尊重，也不过是他的残羹剩炙。星辉之中，明月独皎；日之升矣，星月俱杳。不要凑近使你黯然失色之人，宁可靠拢让你顾盼生辉之辈。正是通过这种方法，马提雅尔[①]诗中的法布拉才显得美丽绝伦、光彩照人，这要归功于她的侍女们其貌不扬、衣冠不整。一个糟糕的伙伴给自己所带来的危害，不见得就比牺牲自己的荣誉而给他人增光的危害更大。当你正走在通往成功的道路上时，可以结交才俊之士；一旦抵达成功的彼岸，你还是结交凡夫俗子吧。

The more he does so the less desirable a companion he is. The more he excels in quality, the more in reputation; he will always play first fiddle and you second. If you get any consideration, it is only his leavings. The moon shines bright alone among the stars; when the sun rises she becomes either invisible or imperceptible. Never join one that eclipses you but rather one who sets you in a brighter light. By this means the cunning Fabula in Martial's verse was able to appear beautiful and brilliant, owing to the ugliness and disorder of her companions. But one should as little imperil oneself by an evil companion as pay honor to another at the cost of one's own credit. When you are on the way to fortune associate with the eminent, when arrived with the mediocre.

---

①马提雅尔(约公元40-120)，古罗马讽刺诗人。

www.buhua.com
2004

不要去填补前人留下的巨大空白

如果非要这么做，你必须要有足够的把握超越前人——仅仅是平分秋色就需要有两倍于他的付出。要让你的后继者导致人们对你的怀念，这样的精心安排才是神来之笔；当心不要让你的前任使你自己黯然失色，这也是精明的策略。要想填补巨大的空白，谈何容易，因为过去看上去总是美妙无比，只和前人打个平手远远不够，因为他拥有占先权。因此，你必须要拥有额外的主张权，才能将他从其所把持的公共舆论中驱逐出去。

BEWARE OF ENTERING WHERE THERE IS A GREAT GAP TO BE FILLED.

But if you do be sure to surpass your predecessor — merely to equal him requires twice his worth. As it is an artful stroke to arrange it so that one's successor shall cause you to be missed, so it is policy to see that our predecessor does not eclipse us. To fill a great gap is difficult, for the past always seems best, and to equal the predecessor is not enough, since he has the right of first possession. You must therefore possess additional claims to oust the other from his hold on public opinion.

www.buhua.com
2004
BUHUA

不要轻易相信，也不要轻易喜爱

心智的成熟，在不轻信上得到了最好的展现。撒谎是司空见惯的事，那么就让信任成为罕见之事吧。轻信盲从之人，很快就会被人小瞧。但是，也大可不必透露出你对他人善意的怀疑。因为这就等于声称那个向你透露消息的人要么是个欺骗者，要么是个被骗者，从而在无礼之上又增加了凌辱。这还不是唯一的不幸。缺乏信任是说谎者的标志，你将会遭受两种失败：既不相信别人，也不被别人相信。就听者而言，推迟做出判断是审慎的，这样，说者就可以求助于最初的信息来源。还有一种类似的轻率，这就是轻易喜爱，因为说谎既可以借助言辞，也可以通过行为，对实际生活来说，后一种欺骗更危险。

DO NOT BELIEVE, OR LIKE, LIGHTLY.

Maturity of mind is best shown in slow belief. Lying is the usual thing, so then let belief be unusual. He that is lightly led away soon falls into contempt. At the same time, there is no necessity to betray your doubts against the good faith of others. For this adds insult to discourtesy, since you make out your informant to be either deceiver or deceived. Nor is this the only evil. Lack of belief is the mark of a liar, who suffers from two failings: he neither believes nor is believed. Suspension of judgement is prudent in a hearer; the speaker can appeal to his original source of information. There is a similar kind of imprudence in liking too easily, for lies may be told by deeds as well as in words, and this deceit is more dangerous for practical life.

控制激情的艺术

如果可能的话，要尽量用审慎的反思抵抗激情的粗暴进击。对真正的智者来说这并不难。控制激情的第一步，就是要认识到自己正处在激情之中。用这种方法，你开始了一场控制自己脾气的战斗，因为一个人必须把自己的激情控制在这样一个精确点上：既必不可少，又不至于过了头。在陷入愤怒和摆脱愤怒的时候，这是艺术中的艺术。你应该知道如何停止以及何时停止才恰到好处——在快速跑步中停下来是最难的。怒发冲冠时保持头脑清醒，是对智慧的巨大考验。每一次激情的过度，都是对理性行为的背离。但借助这种巧妙的策略，理性决不会被违背，也不会超出其自身道德理性的边界。要保持对激情的控制，你必须牢牢控制自己的注意力；能做到这一点，你将会是第一个马背上的智者[①]，没准也是最后一个。

①西班牙谚语云：马背上无智者。

THE ART OF MASTERING YOUR PASSIONS.

If possible, oppose the vulgar advance of passion with prudent reflection. This is not difficult for a truly prudent person. The first step toward mastering a passion is to acknowledge that you are in a passion. By this means you begin the conflict with command over your temper, for one has to regulate one's passion to the exact point that is necessary and no further. This is the art of arts in falling into and getting out of rage. You should know how and when best to come to a stop — and it is most difficult to halt while running double-time. It is a great proof of wisdom to remain clear-sighted during paroxysms of rage. Every excess of passion is a digression from rational conduct. But by this masterful policy reason will never be transgressed, nor pass the bounds of its own moral reason. To keep control of passion one must hold firm the reins of attention; he who can do so will be the first person “wise on horseback”, and probably the last.

择友之道

所谓朋友，只有通过了阅历的考试和命运的测验，他们才能领到毕业证书，考验的不仅是友情，还有眼力。尽管这是生活中最重要的事，但受到的关照却最少。聪明才智给有些人带来朋友，但大多数人还是靠机遇。可以根据一个人的朋友来判断这个人，因为智者和蠢才之间决不会有共鸣。但是，从一个人的社交圈子中得到快乐，也并不能证明亲密友谊的存在：它可能更多的是来自他的同伴的友善愉快，而不是来自对他本人能力的信赖。有些友谊是合法的，而有些友谊则是非法的；后者是因为愉快而存在，前者则是因为他们的观念和动机的丰饶。很少人跟一个人的内心交朋友，大多数是跟他的环境交朋友。一位真心朋友的洞察力，比其他人的善意更有益。因此，要通过选择而不是全凭运气来交朋友。聪明的朋友挡开烦恼，愚蠢的朋友带来烦恼。但不要希望他们有太多的好运，否则你可能会失去他们。

SELECT YOUR FRIENDS.

Only after passing the examination of experience and the test of fortune will they be graduates, not only in affection but in discernment. Though this is the most important thing in life, it is the one least cared for. Intelligence brings friends to some, chance to most. Yet a person is judged by his friends, for there was never sympathy between wise men and fools. At the same time, to find pleasure in a person's society is no proof of close friendship: it may come from the pleasantness of his company more than from trust in his capacity. There are some friendships legitimate, others illicit; the latter for pleasure, the former for their fertility of ideas and motives. Few are the friends of a person's innermost self, most those of his circumstances. The insight of a true friend is more useful than the goodwill of others, therefore gain them by choice, not by chance. A wise friend ward off worries, a foolish one brings them about. But do not wish them too much luck, or you may lose them.

不要看错人

看错人是最糟糕的错误，也是最容易犯的错误。与其在货物品质上受骗，莫如在货物价格上上当。与人周旋，比起与别的事物打交道来，更需要看透内部。识人与识货不同。探测感觉的深度，辨识品格的特征，是一门深奥的哲学，必须像研究书那样深刻地研究人。

DO NOT MAKE MISTAKES ABOUT CHARACTER.

That is the worst and yet easiest error. Better be cheated in the price than in the quality of goods. In dealing with people, more than with other things, it is necessary to look within. To know people is different from knowing things. It is profound philosophy to sound the depths of feeling and distinguish traits of character. People must be studied as deeply as books.

9
9

利用你的朋友

这需要所有的判断技巧。朋友之益，或以其远，或因其近。还有许多朋友，晤谈无益，通信甚佳，盖因为距离使得那些在亲近时无法容忍的缺点不复存在。朋友所带来的实用要超过他们所带来的愉快。美好事物（或者像某些人说的：普遍性存在）所具备的三种品质，朋友都具备：和谐、善良和真诚。一位朋友，就是一切的一切。益友良朋，不可多得，如果你不懂得择友之道，就会变得更少。持友之道，重于交友。选择那些耐久之人作朋友——即使他们起初是新朋友，但总有一天会成为老朋友，这大约会给你一些安慰。最好的朋友当然还是那些腌得很透的朋友，虽然他们或许需要在考验的盐水里浸泡很久。当我们离去时，如果没有朋友，那就是最荒凉的沙漠。友谊，倍增生活的美好，分担生活的不幸。它是疗治不幸的唯一良方，就像新鲜的空气之于心灵。

MAKE USE OF YOUR FRIENDS.

This requires all the art of discretion. Some are good far off, some when near. Many are no good at conversation but excellent as correspondents, for distance removes some failings which are unbearable in close proximity to them. Friends are for use even more than for pleasure, for they have the three qualities of the good, or, as some say, of being in general: unity, goodness, and truth. For a friend is all in all. Few are worthy to be good friends, and even these become fewer because people do not know how to pick them out. Keeping friends is more important than making them. Select those that will wear well — if they are new at first it is some consolation that they will become old. Absolutely the best are those well salted, though they may require soaking in the testing. There is no desert like leaving without friends. Friendship multiplies the good of life and divides the evil. It is the sole remedy against misfortune, like fresh air to the soul.

www.buhua.com
2004

容忍蠢人

聪明人总是缺乏耐心，因为随着知识的增加，他们对蠢行的不耐烦也就随之增加。知识越多越难以满足。据埃皮克提图[①]说，生活最伟大的法则就是要容忍——他认为这抵得上全部智慧的一半。要容忍五花八门的愚蠢，需要足够的耐心。我们常常不得不以最大的耐心容忍那些我们最依赖的人，这是学会自制的一门有用的课程。从耐心中可以生发出平静，而平静，这一珍贵的恩赐乃是现世的福祉。就让那些缺乏忍耐力的人离群索居吧，虽然那样一来，他们将不得不容忍自己。

PUT UP WITH FOOLS.

The wise are always impatient, for he that increases knowledge increases impatience with folly. Much knowledge is difficult to satisfy. The first great rule of life, according to Epictetus, is to put up with things — he valued this as half of all wisdom. To put up with all the varieties of folly would need much patience. We often have to put up with most from those on whom we most depend, which is a useful lesson in self-control. Out of patience comes forth peace, the priceless boon that is the happiness of the world. But let him that has no power of patience then retire within himself, though even there he will have to put up with himself.

---

① 埃皮克提图，公元前一世纪时的希腊斯多噶派哲学家、教师。

www.buhua.com
2004

慎言

说话要小心，跟对手这样是出于谨慎,跟其他人则是为了体面。总有多言的时候,绝无收回的机会。说话要像写遗嘱一样:言辞愈少,争议愈小。要在琐事上锻炼自己，为的是说更重大的事。莫测高深，颇有几分神性的风采。快言快语的人,败得也快。

BE CAREFUL IN SPEAKING.

With your rivals out of prudence, with others for the sake of appearance. There is always time to add a word, never to withdraw one. Talk as if you were making your will: the fewer words the less litigation. In trivial matters exercise yourself for the more weighty matters of speech. Profound secrecy has some of the luster of the divine. He who speaks quickly soon falls of fails.

了解自己珍爱的缺点

最完美的人也有自己珍爱的缺点，要么已经跟它们结了婚，要么在爱着它们。它们常常是才智之士的缺点，智商越高，缺点越大，或者至少是越显眼。拥有这些缺点而不自知的人并不算多，但他珍爱这些缺点，这是双重的不幸，因为这是对可以避免的缺点的无理性偏爱。它们是白璧之玷，它们让拥有者欢喜，也同样让旁观者厌恶。摆脱掉它们，并因此给你的另外一些品质以展示的空间，那将是一件英勇的事情。因为人人都会看到这样的缺点，在审视你的资质的时候，他们会长时间地盯着这个污点看，尽可能使它变得越来越黑，并把你其他的才干扔进阴影里。

KNOW YOUR PET FAULTS.

The most perfect of people has them and is either wedded to them or loves them. They are often faults of intellect, and the greater this is, the greater they are, or at least the more conspicuous. It is not so much that their possessor does not know them, he loves them, which is a double evil because it's an irrational affection for avoidable faults. They are spots on perfection, they displease the onlooker as much as they please the possessor. It is a gallant thing to get clear of them, and so give play to one's other qualities. For all people hit upon such a failing, and on going over your qualifications they will take a long look at this blot and blacken it in as deeply as possible, casting your other talents into the shade.

天才

怎样战胜你的竞争对手和诋毁者

鄙视他们是不够的，虽然这常常是明智的——勇敢行为才是根本之所在。称赞那些贬损你的人，人们就会赞美你，再多也不为过。用你的才华和贡献，去战胜嫉妒并使之痛苦不堪，没有比这更英勇的报复了。对于那些希望你倒霉的人来说，你的每一次成功，都进一步扭紧了他们脖子上的绳套，你的光荣就是竞争对手的地狱。嫉妒者决不止死一次，被嫉妒的人每赢得一次欢呼喝彩，他就死一回。你的不朽美名，就是对手所遭受折磨的衡量标准；一方如果生活在无尽的光荣之中，对方就生活在无穷的痛苦里。名望的号角宣布一方的不朽，也就是宣布另一方的死亡——这是一场被嫉妒所拖延的缓慢死亡。

HOW TO TRIUMPH OVER YOUR RIVALS AND DETRACTORS.

It is not enough to despise them, though this is often wise — a gallant bearing is the essential thing. One cannot praise a person too much who speaks well of them who speaks ill of him. There is no more heroic vengeance than that of talents and services that at once conquer and torment the envious. Every success is a further twist of the cord round the neck of those who wish you ill, and an enemy's glory is the rival's hell. The envious die not once, but as often as the envied wins applause. The immortality of his fame is the measure of the other's torture; the one lives in endless honor, the other in endless pain. The clarion of fame announces immortality to the one and death to the other — the slow death of envy long drawn out.

www.buhua.com
2004

不要因为同情心
而卷入他人的不幸

NEVER-OUT OF SYMPATHY WITH THE UNFORTUNATE-INVOLVE YOURSELF IN THEIR FATE.

一个人的不幸是另一个人的幸运，因为，没有许多人的不幸，一个人不可能幸运。唤起人们的善意正是不幸命运的特点，人们渴望以其无济于事的关切为这次命运之神的打击而给予某些补偿。因万事顺利而受到憎恨的人，碰巧因灭顶之灾而受到关爱。热火朝天的报复突然换成了逶迤而来的同情。然而值得注意的是：厄运是怎样重新洗牌的。有些人总是和倒霉蛋厮混在一起，昨天他还雄心勃勃、喜气洋洋，今天就可怜兮兮地站在不幸者的身边。这样做虽然显得灵魂高贵，但却不是什么处世的智慧。

One person's misfortune is another's luck, for one cannot be lucky without many being unlucky. It is a peculiarity of the unfortunate to arouse people's goodwill, who desire to compensate them for the blows of fortune with their useless favor, and it happens that one who was abhorred by all in prosperity is adored by all in adversity. Vengeance on the wing is exchanged for compassion afoot. Yet it should be noticed how fate shuffles the cards. There are people who always consort with the unlucky, and he that yesterday flew high and happy stands today miserable at their side. That reveals nobility of soul but not worldly wisdom.

www.buhua.com
2004

把稻草扔到空中以测试风向

这样可以发现人们如何理解某件事情，尤其是你对它的成功或者被人接受深表怀疑的事情。这样可以让你对它的最终结果更有把握，并为你提供一次选择的机会：是继续认真干下去，还是及时收手、全身而退。智者利用这种方式测试人们的意图，从而了解他所站的立场。这是远见卓识在寻求、渴望和统驭中的伟大法则。

THROW STRAWS IN THE AIR TO TEST THE WIND.

Find how things will be perceived, especially from those whose reception or success is doubtful. One can thus be assured of its turning out well, and an opportunity is provided for going on in earnest or withdrawing entirely. By trying people's intentions in this way, the wise person knows on what ground he stands. This is the great rule of foresight in asking, in desiring, and in ruling.

发动义战

你可能不得不发动战争，但不要使用毒箭。每个人都必须依据自己的本来面目行事，而不是按照别人所希望的那样去做。在生活的战斗中，勇敢的言行赢得每个人的赞颂；你应该战以致胜，不单单是凭借力量，而且还要靠使用力量的方式。卑鄙的胜利非但不会带来荣耀，反而会带来耻辱。受人尊敬总是占上风。一个值得尊敬的人决不会使用违禁的武器，比如为了刚刚开始的憎恨而终结的友谊；一定不要把信任用于复仇的目的。背信弃义所带来的最轻微的污点，也会玷污你的美名。在受人尊敬的人身上，最轻微的卑劣痕迹也会令人反感。高尚与卑鄙应该远远地分开。要能够自豪地说：已经在世人身上消逝的勇敢、慷慨与忠诚，能够在你自己的心灵中重新发现。

WAGE WAR HONORABLY.

You may be obliged to wage war but not to use poisoned arrows. Everyone must act as he is, not as others would make him to be. Gallantry in the battle of life wins everyone's praise; one should fight so as to conquer, not alone by force but by the way it is used. A mean victory brings no glory, but rather disgrace. Honor always has the upper hand. An honorable person never uses forbidden weapons, such as using a friendship that's ended for the purposes of a hatred just begun; a confidence must never be used for a vengeance. The slightest taint of treason tarnishes one's good name. In people of honor the smallest trace of meanness repels. The noble and the ignoble should be miles apart. Be able to boast that is gallantry, generosity, and fidelity were lost in the world people would be able to rediscover them in your own heart.

区分空谈家和实干家

这种区别很重要，在朋友、个人和职业方面都是如此，这些全都有着形形色色的类型。糟糕的言辞即使没有糟糕的行动，也已经够糟糕的了；而良好的言辞加上糟糕的行动则更糟糕。你不可能拿言辞当饭吃，那是风；也不可能用优雅来果腹，那只不过是文雅的欺骗。用镜子来捕鸟，是空想的陷阱。想在风一般的言辞中取得报酬的人，只能是白费力气。言辞应该是工作的典当物，像当票一样也有其市场价格。只长叶子不结果的树，通常没有核——从它们的本质可知：它们除了乘凉，别无用处。

DISTINGUISH PEOPLE OF WORDS FROM PEOPLE OF DEEDS.

Discrimination is important, as in the case of friends, persons, and employments, which all have many varieties. Bad words even without bad deeds are bad enough; good words with bad deeds are worse. One cannot dine off words, which are wind, nor off politeness, which is but polite deceit. To catch birds with a mirror is the ideal snare. It is the vain alone who take their wages in windy words. Words should be the pledges of work, and, like pawntickets, have their market price. Trees that bear leaves but not fruit usually have no core — know them for what they are, of no use except for shade.

懂得如何依靠自己

在危急关头，没有比勇敢的心灵更好的伙伴了；如果它变得软弱，就必须通过相邻的部位来增强它。烦恼因为坚持自我的人而逐渐消失。千万不要向厄运低头，否则它会变得无法忍受。许多人在困境中不能自救，并因为不懂的如何承受它们而让它们变本加厉。了解自己的人，懂得如何让自己的弱处得以增强。智者征服一切，甚至包括命运。

KNOW HOW TO RELY ON YOURSELF.

In great crises there is no better companion than a bold heart, and if it becomes weak it must be strengthened from the neighboring parts. Worries dies away for the person who asserts himself. One must not surrender to misfortune or else it would become intolerable. Many people do not help themselves in their troubles and double their weight by not knowing how to bear them. He that knows himself knows how to strengthen his weakness, and the wise person conquers everything, even the stars in their courses.

不要纵容愚蠢的怪癖

像虚荣自负、专横傲慢、自我膨胀、不可信赖、反复无常、顽固偏执、耽于幻想、矫揉造作、异想天开、追新逐奇、自相矛盾以及形形色色的剑走偏锋——这些全都是粗鲁无礼的畸形。一切心智的畸形都比身体畸形更可憎，因为它亵渎了更高贵的美。然而，对于心智的这种彻底混乱，谁又能帮得了呢？缺乏自制力的地方，没有他人施以援手的空间。这种人，并不注意别人真实的嘲弄，却盲目追求虚假的希望：希望赢得想象中的欢呼喝彩。

DO NOT INDULGE IN THE ECCENTRICITIES OF FOLLY.

Like vanity, presumptuousness, egotism, untrustworthiness, capriciousness, obstinacy, fancifulness, theatricalism, whimsy, inquisitivness, contradiction, and all forms of one-sidedness — they are all monstrosities of impertinence. All deformity of mind is more obnoxious than that of the body, because it violates a higher beauty. Yet who can assist such a complete confusion of mind? Where self-control is wanting, there is no room for others' guidance. Instead of paying attention to other people's real derision, people of this kind blind themselves with the false hope of imaginary applause.

避免失手一次，胜过击中百回

烈日当空之时，谁也不会瞧上一眼；一旦出现日蚀，人人都会盯着看。人们对好事置若罔闻，坏事却总是津津乐道。好事不出门，坏事传千里。许多人不为人知，直到他们撒手人寰。一个人毕生的功绩，加在一起也不够擦除一个小小的污点。因此，尽量避免犯错，要知道恶意总是在注意每一个失误，而不在乎任何一次成功。

BE MORE CAREFUL NOT TO MISS ONCE THAN TO HIT A HUNDRED TIMES.

No one looks at the blazing sun, but all gaze when it is eclipsed. The common talk does not reckon what goes right but what goes wrong. Evil news carries farther than any applause. Many people are not known to the world till they have left it. All the exploits of a person taken together are not enough to wipe out a single small blemish. Avoid therefore falling into error, knowing that ill will notices every error and no success.

www.buhua.com
2004

凡事留一手

这是维持你的重要性的可靠途径。你应该时刻准备同时用上你所有的才能和力量。即使在知识领域,也应该有后卫,这样你的资源就会倍增。当你害怕被击败的时候,总要有某些东西可供凭藉。后备部队是一支比进攻主力更重要的力量,因为它以勇敢和荣誉为特征。审慎之人总是在有了安全保证的前提下才开始行动。在这个问题上,我们不妨相信这样一个有趣的悖论:一半多于全部。

IN ALL THINGS KEEP SOMETHING IN RESERVE.

This is a sure way of keeping up your importance. A person should employ all his capacity and power at once and on every occasion. Even in knowledge there should be a rearguard so that your resources are doubled. One must always have something to resort to when there is fear of a defeat. The reserve is of more importance than the attacking force, for it is distinguished by valor and reputation. Prudence always sets to work with assurance of safety. In this matter the piquant paradox holds true: the half is more than the whole.

2004

不要滥用你的影响力

像朋友这种重要东西，是为重要时刻准备的。一个人不能为了琐碎之事而随意利用伟大的信任，那是在滥用友爱。紧急抛锚应该留作最后的手段。如果你为了小结果而把大力量都用完了，那么你留下什么应对将来呢？如今，没有什么东西比保护者更加宝贵，也没有什么东西比友爱更有价值。它既能创造、也能毁灭整个世界。它甚至既能赐予、也能剥夺你的智慧。天性和名声有利于智者，幸运之神通常也对此艳羡不已。是故，保存有力者的友善关爱，比保存有形财产更为重要。

DO NOT WASTE INFLUENCE.

The great as friends are for great occasions. One should not make use of great confidence for little things, for that wastes a favor. The emergency anchor should be reserved for the last resort. If you use up the great for little ends what remain afterward? Nothing is more valuable than a protector and nothing costs more nowadays than a favor. It can make or unmake a whole world. It can even support your wits or take them away. As nature and fame are favorable to the wise, so luck is generally envious of them. It is therefore more important to keep the favor of the mighty than goods and chattels.

www.buhua.com
2004

了解你的幸运之星

没有幸运之星的人最无助。如果他不幸，那是因为他对自己的幸运之星懵然无知。有些人身居高位，深受王公贵戚的宠信，却既不知其然也不知其所以然，只是好运本身以优厚的条件给了他们这样的恩惠，几乎不需要他们出力襄助。有些人则以他们的聪明赢得了厚爱。有的人，被一个国家接纳，而遭另一个国家拒斥；在一座城市大受欢迎，而在另一座城市倍遭冷落；在一个职位上吉星高照，而在另一个职位上倒霉透顶；所有这些，条件并无不同，结果却相去甚远。幸运女神何时洗牌、如何洗牌，全凭她的意志。让每个人既了解自己的天资才能，也了解自己的幸运之星，是输是赢，皆在于此。遵循你的幸运之星的指引吧，即使它的邻居（没准是北极星吧）以雷鸣般的声音召唤我们随它而去；认准它，帮助它，不要把它错认作别的东西，否则你将找不着北。

KNOW YOUR RULING STAR.

No one is so helpless as not to have a ruling star; if he is unlucky, that is because he does know it. Some stand high in the favor of princes and potentates without knowing why or wherefore, except that good luck itself has granted them favor on easy terms, merely requiring them to aid it with a little exertion. Others find favor with the wise. One person is better received by one nation than another, or is more welcome in one city than another. He finds more luck in one office or position than another, and all this though his qualifications are equal or even identical. Luck shuffles the cards how and when she will. Let each person know his luck as well as his talents, for on this depends whether he loses or wins. Follow your guiding star and help it without mistaking it for any other, for that would be to miss the north, though its neighbor (or polestar) calls us to it with a voice of thunder.

www.buhua.com
2004

别做玻璃人

DO NOT BE MADE OF GLASS IN YOUR RELATIONS WITH OTHERS, STILL LESS IN FRIENDSHIP.

在与他人的交往中，不要做玻璃人，在友谊中则更是如此。有些人很容易破碎，由此显示他们缺少坚固性。他们总是沉湎于假想的冒犯，沉湎于其他人压制的意向。他们的感觉甚至比眼睛本身还要敏感，万万碰不得，无论你是嬉皮笑脸还是一本正经。尘埃之微，也会冒犯他们，这就更不用说桁梁了。那些与他们交往的人，必须拿出最大的细心来应付他们，小心对待他们敏感的神经，注视他们的行为举止，因为最轻微的怠慢也会惹恼他们。他们主要是以自我为中心，是自己情绪的奴隶，为此他们把所有事情都抛到一边。他们是琐事的崇拜者。另一方面，真正的情人，其性情几乎就像钻石一样：坚固而恒久。

Some break very easily, and thereby show their want of consistency. They attribute to themselves imaginary offences and to others oppressive intentions. Their feelings are even more sensitive than the eye itself and must not be touched in jest or in earnest. Motes offend them; they need not wait for beams. Those who consort with them must treat them with the greatest delicacy, have regard to their sensitiveness, and watch their demeanor, since the slightest slight arouses their annoyance. They are mostly very egoistic, slaves of their moods, for the sake of which they cast everything aside. They are worshippers of little nothings. On the other hand, the disposition of the true lover is almost diamond-like: hard and everlasting.

不要活得太匆忙

懂得怎样将事情分门别类，也就懂得如何享受其中的乐趣。许多人好运虽已用完，生命却未耗尽，等到他们发现自己跑过了头，才会想到迷途知返。他们是生活马车的驭者，心中愿望急迫，自然快马加鞭。他们一天的狼吞虎咽，整个一生也未必消化得了。他们总是超前享受，寅吃卯粮，因为自己的匆忙，凡事总是完成得太快。即使在探索知识的时候，也要不紧不慢，这样才使我们不至于学得太多，懂得太少。寻常日子多，快乐光阴少。工作宜快，享乐宜缓，因为人们发现：工作总用快乐做结，享乐皆以叹憾告终。

DO NOT LIVE IN A HURRY.

To know how to separate things is to know how to enjoy them. Many people finish their fortune sooner than their life. They run through pleasures without enjoying them, and would like to go back when they find they have overrun the mark. Postilions of life, they increase the ordinary pace of life by the hurry of their own calling. They devour more in one day than they can digest in a whole lifetime; they live in advance of pleasures, eat up the years beforehand, and by their hurry get through everything too soon. Even in the search for knowledge there should be moderation, lest we learn things better left unknown. We have more days to live through than pleasures. Be slow in enjoyment, quick at work, for people see work ended with pleasure, pleasures ended with regret.

www.buhua.com
2004

做个坚实的人

一个坚实的人，在那些不坚实的事物中得不到满足。根基不牢固的显赫，是一种可怜的显赫。并非所有东西都是看上去的那个样子。有些是欺骗的源头——它们通过狂想而受孕，从而产生出欺诈。另一些则很像它们，以至于它们更乐于谎言（因为承诺很多），而不是事实（因为践履甚少）。但是到最后，这些异想天开的妄想终会带来糟糕的结局，因为它们没有坚实的基础。只有真理才能带来真正的名声，只有事实才有实在的利益。一个欺骗需要更多别的欺骗来掩盖，整个房屋因此成为空中楼阁，很快就会坍塌。没有根基的东西注定短命。它们许诺太多，以至于很难被人信任：总是需要验证的东西不可能是真的。

A SOLID PERSON.

One who is finds no satisfaction in those that are not. It is a pitiable eminence that is not well founded. Not all are those that seem to be so. Some are sources of deceit — impregnated by chimeras, they give births to impositions. Others are like them so much that they take more pleasure in a lie (because it promises much) than in the truth (because it performs little). But in the end these caprices come to a bad end, for they have no solid foundation. Only truth can give true reputation; only reality can be of real profit. One deceit needs many others, and so the whole house is built in the air and must soon come to the ground. Unfounded things never reach old age. They promise too much to be much trusted: that cannot be true that proves too much.

名人

要有知识，或者认识有知识的人

如果不拥有聪明才智（无论是自己的还是别人的），就不可能有真正的生活。但许多人认识不到自己不懂什么，而另一些人则在自己一无所知的时候自认为无所不知。智力上的缺点是不可救药的，因为无知者亦无自知之明，因此不可能去寻求自己所缺少的东西。如果不自认为聪明的话，许多人本来会很聪明。因此，尽管富有智慧的贤哲弥足珍贵，但却很少被人们所用。求教于人无损于你的伟大，也不会显得你无能。相反，寻求他人的忠告，会证明你的明智。如果你不想招致失败，不妨就教于智者。

HAVE KNOWLEDGE, OR KNOW THOSE WHO DO.

Without intelligence, either one's own or another's, true life is impossible. But many do not know that they do not know, and many think they know when they know nothing. Failings of the intelligence are incorrigible, since those who do not know, do not know themselves, and cannot therefore seek what they lack. Many would be wise if they did not think themselves wise. Thus it happens that though the oracles of wisdom are precious, they are rarely used. To seek advice does not lessen greatness or argue incapacity. On the contrary, to ask advice proves you well advised. Take counsel with reason if you do not wish to court defeat.

避免太亲近他人

既不要太亲近他人，也别让他人太亲近你。否则，你就会失去你的影响力所赋予你的优势，并且失去他人对你的尊敬。星星总是拒人于千里之外，所以才能保持它们的光亮。神性需要庄重，亲昵导致轻慢。在人类事务中，一个人显露的越多，拥有的就越少，因为在公开的交流中，你把那些被矜持所掩盖的缺点都暴露无遗。亲昵从来就不是什么得体之事：施之于上则十分危险，施之于下又有失身份，施之于群氓则尤为不可，他们因彻头彻尾的愚蠢而傲慢无礼——他们错将你的好意误认作你觉得需要他们。亲昵近于粗俗。

AVOID BEING TOO FAMILIAR WITH OTHERS.

Nor should you permit other to be too familiar with you. He that is too familiar loses any superiority his influence gives him and so loses respect. The stars keep their brilliance by not making themselves common. The divine demands decorum. Every familiarity breeds contempt. In human affairs, the more a person shows the less he has, for in open communication you communicate the failings that reserve might keep under cover. Familiarity is never desirable: with superiors because it is dangerous, with inferiors because it is unbecoming, least of all with the common herd, who become insolent from sheer folly — they mistake favor shown them for need felt of them. Familiarity verges on vulgarity.

2004
www.buhua.com

相信自己的内心

决不要拒绝听取内心的声音。它是一种内生性神谕，常常能预言最重要的事情。许多人因为恐惧自己的内心而走向毁灭，但如果没有找到更好的疗救之法，恐惧又有何益？大自然赋予了许多人一颗如此真诚的心灵，它总是对即将来临的不幸发出警告，并努力避免不幸的结果。自寻灾祸实属不智，除非你是要试图征服它们。

TRUST YOUR HEART. ESPECIALLY WHEN IT HAS BEEN PROVED.

Never deny it a hearing. It is a kind of house oracle that often foretells things most important. Many have perished because they feared their own heart, but of what use is it to fear it without finding a better remedy? Many are endowed by nature with a heart so true that it always warns them of misfortune and wards off its effects. It is unwise to seek evils, unless you seek to conquer them.

www.buhua.com
2004

缄默是能力的封印

一颗没有秘密的心灵，就是一封打开的书信。有坚实基础的地方，秘密能得到深刻的保持——有些华而不实的地窖，一些重要的东西可能就深藏其中。缄默来自自制，在这一点上控制自己，就是真正的胜利。你必须为你说出的每一句话支付赎金。智慧的安全在于内心的节制。缄默所要冒的风险，就潜藏在他人的反复盘问当中，潜藏在对透露出的秘密使用反驳当中，潜藏在讽刺的投枪当中。为了避开这些，审慎之人变得比从前更缄默。必做之事不必说，必说之事不必做。

RETICENCE IS THE SEAL OF CAPACITY.

A heart without a secret is an open letter. Where there is a solid foundation secrets can be kept profound — there are specious cellars where important things may be hidden. Reticence springs from self-control and to control oneself in this is a true triumph. You must pay ransom to each you tell. The security of wisdom consists of inner temperance. The risk that reticence runs lies in the cross-questioning of others, in the use of contradiction to worm out secrets, in the darts of irony. To avoid there the prudent become more reticent than ever. What must be done need not be said, and what must be said need not be done.

决不要把敌人引到他不得不做的事情上

蠢人从不做智者所英明决断的事情，因为他不会用恰当的手段去贯彻。小心谨慎的人，则更不会遵循由别人制订的计划，哪怕是已经实现的计划。你必须从双方的观点来探讨问题——把它放到双方的立场上反复考量。得出的判断会有所不同。如果你还没有做出决定，与其留意可能之事，不如专注可行之事。

NEVER GUIDE THE ENEMY TO WHAT HE HAS TO DO.

The fool never does what the wise judge wise, because he does not follow up with suitable means. He that is discreet follows still less a plan laid out, or even carried out, by another. One has to discuss matters from both points of view — turn it over on both sides. Judgements vary. Let him that has not decided attend rather to what is possible than what is probable.

人体炸弹!!

说出真相,但不是全部真相

没有什么东西比真相需要更加小心的了——它是心灵的手术刀。要说出真相,同样也要隐瞒真相。一句谎言,可以使诚实的美名毁于一旦。欺骗被视为背叛,而说谎者则被视为叛徒——这更糟。然而,并不是所有真相都能言之于口,有时也必须藏之于心,有些是为我们自己着想,有些则是为他人考虑。

THE TRUTH, BUT NOT THE WHOLE TRUTH.

Nothing demands more caution than the truth — it is the lancet of the heart. It requires as much to tell the truth as to conceal it. A single lie destroys a whole reputation for integrity. The deceit is regarded as treason and the deceiver as a traitor, which is worse. Yet not all truths can be spoken, some for our own sake, others for the sake of others.

www.buhua.com
2004

凡事大胆一点

这是一项重要的审慎之道。你必须有节制地看待他人，这样就不至于把他们看得过高从而害怕他们。想象不应该屈从于心智。许多人在你亲自了解他们之前看上去很了不起，然而一旦打过交道，更多的是让你大失所望，而不是肃然起敬。没有谁能超越人性的狭隘——人人都有缺点，要么是心灵上的，要么是头脑上的。显赫的职位给人以表面的权威，但很少伴随着个人的能力，因为命运之神常常以占据高位者的卑下来平衡权位的高度。想象总是跑的太快，它总是用更鲜艳的色彩而不是按照真实的样子来描绘事物。它不是按照事物的本来面目而是依据自己希望的样子来看待事物。尽管从前大失所望，但专注很快就会纠正一切。然而，如果说智者不该胆怯的话，那么愚人也不该鲁莽。如果说自立对无知者有所助益的话，那么它对智勇之士的帮助不是更大么?

A GRAIN OF BOLDNESS IN EVERYTHING.

This is an important piece of prudence. You must moderate your opinion of others so that you may not think so high of them as to fear them. The imagination should never yield to the heart. Many appear great till you know them personally, and then dealing with them does more to raise disillusion than esteem. No one oversteps the narrow bounds of humanity — all have their weaknesses either in heart or head. Dignity gives apparent authority, which is rarely accompanied by personal power, for fortune often redresses the height of office by the inferiority of the holder. The imagination always jumps too soon and paints things in brighter colors than the real. It thinks things not as they are but as it wishes them to be. Attentiveness — though disillusioned in the past — soon corrects all that. Yet if wisdom should not be timorous, neither should folly be rash. And if self-reliance helps the ignorant, how much more the brave and wise?

不要固执己见

每个蠢人都固执，每个固执的人都是蠢人;观点愈错，持之愈坚。即使在确凿无疑的情况下，适当的让步亦是良策。我们抱持这种观点的理由，别人不可能视而不见，我们让步时的谦恭也会被人认可。我们的固执却让我们得不偿失——那与其说是坚持真理，还不如说是粗暴无理。有些榆木脑袋很难扭转，固执己见加上异想天开就是令人生厌的愚蠢。坚定不移的，应该是意志，而不是看法。然而也有例外的情况，你会两次犯错，一是判断上的错误，二是执行上的错误。

DO NOT HOLD YOUR VIEWS TOO FIRMLY.

Every fool is firmly convinced, and everyone fully persuaded is a fool; the more erroneous his judgement the more firmly he holds it. Even in cases of obvious certainty, it is fine to yield. Our reasons for holding the view cannot escape notice, our courtesy in yielding will be recognized. Our obstinacy loses more than our victory gains — that is not to champion truth but rather rudeness. There are some heads of iron most difficult to turn, and add caprice to obstinacy and the sum is a wearisome fool. Steadfastness should be for the will, not for the mind. Yet there are exceptions where one would fail twice, owning oneself wrong both in judgement and in the execution of it.

不要拘泥礼节

即使是在君王的身上，这种矫揉造作也会以怪癖而闻名。拘泥礼节就是令人生厌，更有甚者是整个民族都有这种怪癖。蠢人的装束就是用这种东西编织而成的。这样的民族，崇拜的是自己的体面，而很少显示这种体面的正当性，因为他们担心最细微的事情也能毁掉体面。希望得到人们的尊重并不错，但并非要让人认为你是个礼节大师。事实上，做事而不拘泥礼节需要非凡的才能。既不要受礼节的影响，也不要轻视礼节——在这样的小事上显得很了不起的人，不可能有什么大的了不起。

DO NOT STAND ON CEREMONY.

Even in kings this affectation is renowned for eccentricity. To be punctilious is to be a bore, yet whole nations have this peculiarity. The garb of folly is woven out of such things. Such folk are worshippers of their own dignity, yet show how little it is justified since they fear that the least thing can destroy it. It is right to demand respect, but not to be considered a master of ceremonies. Yet it is true that in order to do without ceremonies one must possess supreme qualities. Neither affect nor despise etiquette — he cannot be great who is great at such little things.

决不要拿你的名声孤注一掷

如果输掉了老本，损失将无法挽回。很有可能发生是：你一掷而败（特别是第一次）。环境并非总是有利，因此人们总是说："风水轮流转。"你的第二次努力总是和第一次有关，如果初战告捷，则可以为第二次打点底子，万一失败，第二次还有挽回的希望。总归要凭借更好的手段，求助于更多的资源。事情的成败得失，取决于各种各样的机会。这也就是为什么令人满意的成功如此罕见的缘故。

NEVER STAKE YOUR CREDIT ON A SINGLE CAST OF THE DICE.

If it miscarries the damage is irreparable. It may easily happen that you might fail once, especially at first. Circumstances are not always favorable, hence they say, "Every dog has his day." Always connect your second attempt with your first, because whether it succeeds or fails the first will redeem the second. Always have resort to better means and appeal to more resources. Things depend on all sorts of chances. That is why the satisfaction of success is so rare.

www.buhua.com
2004

要认识人的缺点，无论他位置多高

诚实正直，能够发现隐藏在绫罗绸缎甚或金顶王冠之下的丑恶，但尽管如此，却不能隐藏其自身的品格。奴隶不可能用主人的高贵来掩饰自己的卑劣。丑恶也可以身居高位，但即便如此，它也总是卑下的。人们可以看到，许多大人物都有大缺点，然而他们却看不到，他并不是因为这些缺点而伟大。伟人的榜样是如此堂皇其表，以至于甚至掩盖了邪恶，直至对那些阿谀奉承者产生很大的影响，使得他们没有注意到：掩盖在堂皇外表之下的，正是他们所憎恶的低级阶层身上所拥有的那些东西。

RECOGNIZE FAULTS, HOWEVER HIGHLY PLACED.

Integrity can discover vice when clothed in brocade or even crowned with gold, but will not be able to hide its own character for all that. Slavery does not lose its vileness because it is disguised by the nobility of its lord and master. Vices may stand in a high place, but are low for all that. People may see that many a great person has great faults, yet they do not see that he is not great because of them. The example of the great is so specious that it even glosses over viciousness, until it may so affect those who flatter it that they do not notice that what they gloss over in the great they abominate in the lower classed.

令人愉快的事情自己做，叫人讨厌的事情让别人做

前一种行为让你赢得好感，后一种做法让你避免敌意。伟人更乐于施惠于人，而不是受人恩惠——这是他慷慨天性的特权。如果自己不曾遭受由同情或懊悔所带来的痛苦，恐怕也不那么容易让别人遭受痛苦。身居高位者，其赖以开展工作的手段不外乎两种：赏与罚；前者须得亲自操办，后者不妨他人代劳。让某人面对不满、憎恨和诽谤的种种兵器，他就会被你所控制。因为，乌合之众的恼怒就像疯狗：不知道自己痛苦的根源之所在，转而去嘶咬鞭子本身，虽然鞭子并非真正的罪犯，却不得不代人受过。

DO PLEASANT THINGS YOURSELF, UNPLEASANT THINGS THROUGH OTHERS.

By the one course you gain goodwill, by the other you avoid hatred. A great person takes more pleasure in doing a favor than in receiving one — it is the privilege of his generous nature. One cannot easily cause pain to another without suffering pain either from sympathy or from remorse. In a high position one can only work by means of rewards and punishment, so grant the first yourself, inflict the other through others. Have someone against whom the weapons of discontent, hatred, and slander may be directed. For the rage of the mob is like that of a dog: missing the cause of its pain it turns to bite the whip itself and, though this is not the real culprit, it has to pay the penalty.

www.buhua.com
2004

做赞美的传递者

这会增加人们对我们的良好品味的信任，因为它表明，我们已经在别的地方学会了懂得什么是卓越，并因此懂得如何在面前的同伴中珍视它。它为交谈提供了材料，为仿效提供了原型，并鼓励那些值得赞赏的努力。此外，这也是在以一种十分微妙的方式向我们面前的卓越品质表示敬意。另一些人则恰恰相反，他们说起话来冷嘲热讽，通过贬低不在场的人来讨好在场的人。这或许可以为他们招徕浅薄之徒，这些人没有注意到在人背后说坏话是多么狡诈。许多人总是热衷于高估今天的凡夫俗子，而贬低过去的丰功伟业。让审慎者洞穿这些阴险狡诈的把戏吧，既不因此人的夸张而沮丧，也不因彼人的吹捧而自负；要知道他们是在用不同的方法、以同样的方式跟身边的同伴说话。

BE THE BEARER OF PRAISE.

This increases our credit for good taste, since it shows that we have learned elsewhere to know what is excellent and hence how to prize it in the present company. It gives material for conversation and for imitation and encourages praiseworthy exertions. Besides, this does homage in a very delicate way to the excellences before us. Others do the opposite, they accompany their talk with a sneer, and fancy they flatter those present by belittling the absent. This may serve them with superficial people, who do not notice how cunning it is to speak ill of everyone to everyone else. Many pursue the plan of valuing more highly the mediocrities of the day than the most distinguished exploits of the past. Let the cautious penetrate through these subtleties, and let him not be dismayed by the exaggerations of the one or made overconfident by the flatteries of the other; knowing that both act in the same way by different methods, adapting their talk to the company they are in.

一点都不像猪！
是是是！一看
就是自己人！

利用他人的欲求

一个人的欲求越大，压力也就越大。哲学家说，匮乏就是乌有；而政治家说，匮乏无所不包，他们是对的。许多人拿别人的需求做成自己实现目的的阶梯。他们利用这个机会，指出满足需求的困难，从而挑起别人的欲望。欲望的活力，比起占有的惰性，能够带来更多的期望。欲望的激情，随着阻力的增长而不断增长。微妙之处，就在于满足欲望的同时又维持他人对你的依赖。

UTILIZE ANOTHER'S WANTS.

The greater his wants the greater the turn of the screw. Philosophers say privation is non-existent, but statesmen say it is all-embracing, and they are right. Many make ladders to attain their ends out of the wants of others. They make use of the opportunity and tantalize the appetite by pointing out the difficulty of satisfaction. The energy of desire promises more than the inertia of possession. The passion of desire increases with every increase of opposition. It is a subtle point to satisfy the desire and yet preserve the dependence.

老婆啊，海里根本没鱼，只能给你抓这个了！

在一切事情中寻找慰藉

即使是毫无用处的东西也能在不朽中找到安慰。任何烦恼都有其补偿。傻人有傻福,丑人的好运众所周知。价值不大的人,寿命更长——破碎的玻璃决不会被打碎,倒是它的耐久性让你烦恼。看来,命运之神似乎也嫉贤妒能,于是让庸人长寿,令英才早逝。肩负重任的人很快就会遭遇不幸,而无关紧要之辈总是活得无忧无虑:一种情况是看上去是这样,另一种情况是确实如此。在那些不幸的人看来,幸运之神和死神都已经把自己给忘了。

FIND CONSOLATION IN ALL THINGS.

Even the useless may find it in being immortal. No trouble without compensation. Fools are held to be lucky, and the good luck of the ugly is proverbial. Be worth little and you will live long — it is the cracked glass that never gets broken, but worries one with its durability. It seems that fortune envies the great, so it equalizes things by giving long life to the useless, a short one to the important. Those who bear the burden come soon to grief, while those who are of no importance live on and on: in one case it appears so, in the other it is so. The unlucky thinks he has been forgotten by both death and fortune.

丑，可是
很软乎。

别想从彬彬有礼中得到报偿

这是一种欺骗。有些人不需要灵丹妙药来制作他们的魔剂，他们只需凭借举手礼的优雅就足以让傻瓜神魂颠倒。他们拥有的是“优雅银行”，支付的只是一些随风而逝的甜言蜜语。允诺一切就是什么也没允诺——允诺是傻瓜的陷阱。真正的谦恭，是职责的履行；而虚假的谦恭，尤其是毫无用处的谦恭，则是欺骗。这与其说是尊敬，不如说是巴结权势的手段。他们尊敬的对象，并不是人，而是他的钱；恭维，也并非献给他们所认可的品格，而是献给他们渴望得到的好处。

DO NOT TAKE PAYMENT IN POLITENESS.

This is a kind of fraud. Some do not need exotic herbs for their magic potion, for they can enchant fools by the grace of their salute. Theirs is the Bank of Elegance, and they pay with the wind of fine words. To promise everything is to promise nothing — promises are the pitfalls of fools. The true courtesy is performance of duty; the spurious, and especially the useless, is deceit. It is not respect but rather a means to power. Obeisance is paid not to the man but to his means, and compliments are offered not to the qualities that are recognized but to the advantages that are desired.

平和即长寿

要想自己活，就得让别人活。心平气和的人不仅仅过生活，而且还主宰生活。多听，多看，少说。白天与世无争，晚上无梦而眠。漫长而快乐的一生，足以抵得上两辈子——这就是平和带来的结果。对与己无关的事情毫不在乎，你就应有尽有。最反常的莫过于事事关心。与己无关的事牵肠挂肚，而切身相关的事却毫不上心，这二者同样愚蠢。

A PEACEFUL LIFE IS A LONG LIFE.

To live, let live. Peacemakers not only live, they rule life. Hear, see, and be silent. A day without dispute brings sleep without dreams. Long life and a pleasant one is life enough for two — that is the fruit of peace. He has all that makes nothing of what is nothing to him. There is no greater perversity than to take everything to heart. There is equal folly in troubling our heart about what does not concern us and in not taking to heart what does.

提防那些始于利他、终于利己的人

WATCH OUT FOR PEOPLE WHO BEGIN WITH ANOTHER'S CONCERNS TO END WITH THEIR OWN.

对于狡诈，警觉是唯一的保护。要关注他们的意图。许多人很善于让别人帮自己做事，除非你掌握了他们的动机，否则你随时都可能被迫为他们火中取栗，而伤及自己的手指。

Watchfulness is the only guard against cunning. Be intent on their intention. Many succeed in making others do their own affairs, and unless you possess the key to their motives you may at any moment be forced to take their chestnuts out of the fire to the damage of your own fingers.

快披上！别冻着！

理性看待自己和自己的事业

HAVE REASONABLE VIEWS OF YOURSELF AND OF YOUR AFFAIRS.

如果你处在人生的起点，这一条尤其真切。每个人都自视甚高，底子寒薄的人更是如此。人人都梦想着自己的好运，认为自己是个旷世奇才。心比天高，命比纸薄。当幻灭伴随现实一同到来的时候，像这样不切实际的空想，只不过是充当了烦恼的源泉。明智之士早就预见了这样的错误。或许，他也总是希望最好的结果，但是，他却一直在作最坏的预期，这样一来，他才会平静接受即将到来的一切。真的，在人生最初的起点，把目标定在一个你可以达到的高度，而不要高到你遥不可及的程度，这才是明智之举。对期望值进行这样的修正是必要的，因为在未曾经历之前，期望值肯定总是太高。疗救愚蠢，最有效的灵丹妙药是审慎。如果你了解自己才能和位置的真实范围，你就能使理想和现实相协调。

This is especially true in the beginning of life. Everyone has a high opinion of himself, especially those who have least ground for it. Everyone dreams of his good luck and thinks himself a marvel. Hope gives rise to extravagant promises that experience does not fulfill. Such idle imaginations merely serve as a wellspring of annoyance when disillusion comes with the true reality. The wise man anticipates such errors. He may always hope for the best, but he always expects the worst, so as to receive what comes with equanimity. True, it is wise to aim high so as to hit your mark, but not so high that you miss your mission at the very beginning of life. This correction of expectations is necessary because before experience comes, expectation is sure to soar too high. The best panacea against folly is prudence. If you know the true sphere of your activity and position, you can reconcile ideals with reality.

www.buhua.com
2004

懂得如何欣赏他人

人人皆可为师，没有人杰出到不可超越的程度。懂得如何利用别人，是一门很有用的知识。聪明之人欣赏每一个人，因为他们看到了每个人身上的优点，也懂得做好任何一件事情有多难。蠢人总是贬低他人，不识好，专挑歹。

KNOW HOW TO APPRECIATE.

There is no one who cannot teach somebody something, and there is no one so excellent that he cannot be excelled. To know how to make use of everyone is useful knowledge. Wise men appreciate everyone, for they see the good in each and know how hard it is to make anything good. Fools depreciate everyone, not recognizing the good and selecting the bad.

哇！倒着看你真
不像鬼哎！

别把蠢人扛在自己背上

见到蠢人而辨认不出来的人，自己就是个蠢人，而认出来了却不躲开的人，则更是蠢人。他们是危险的同伴和毁灭性的知己。即使暂时他们自己谨慎小心，而别人也注意与他们保持距离，但最终他们肯定会有更加愚蠢的言行，这些他们已经雪藏了许久。对于不接受他们的信任的人，他们也得不到他的信任。他们是最不幸的，这是蠢人的报应，他们不得不因为这样那样的事情而付出代价。关于蠢人，只有一件事情还不算太糟，那就是，尽管他们对智者来说并没有什么用处，但作为警示信号或者路标他们还是很不错的。

DO NOT CARRY FOOLS ON YOUR BACK.

He that does not know a fool when he sees one is one himself, still more he that knows him but will not keep clear of him. They are dangerous company and ruinous confidants. Even though their own caution and others' care keeps them in bounds for a time, still at length they are sure to do or to say some foolishness that is all the greater for being kept so long in stock. They cannot help another's credit who have none of their own. They are most unlucky, which is the nemesis of fools, and they have to pay for one thing or the other. There is only one thing that is not so bad about them, and this is that though they can be of no use to the wise, they are good as warning signs or as signposts.

懂得如何异地而居

对有些民族，你必须越过他们的边境才能让人认识到你自身的价值，尤其是当你处在重要职位上的时候。他们的本土就是伟大天才的后母；嫉妒在其出生的土壤中茂盛生长，他们记住的，是你渺小的起点，而不是你成就的伟业。一根走遍世界的针被人们所珍赏，一片来自远方的彩绘玻璃在价值上胜过钻石。每一件外来事物都受到人们的敬重，这部分地是因为它来自远方，部分地是因为它是现成的、已经完善了的。我们都见识过，一个曾经是乡里笑柄的人，如今成了全世界的奇迹，受到他的同胞们和外国人的尊敬——后者是因为他们来自远方，前者则是因为他们是从远处观看。祭坛上的木制雕像，决不会让那个知道它本是花园里的一根树干的人感到敬畏。

KNOW HOW TO TRANSPLANT YOURSELF.

There are nations with whom one must cross their borders to make one's value felt, especially when in great posts. Their native land is always the stepmother to great talents; envy flourishes there on its native soil and they remember one's small beginnings rather than the greatness one has reached. A needle is appreciated that comes from one end of the world to the other, and a piece of painted glass might outvie the diamond in value if it comes from afar. Everything foreign is respected, partly because it comes from afar partly because it is ready made and perfect. We have seen a person once the laughingstock of their village and now the wonder of the whole world, honored by their fellow countrymen and by foreigners — by the latter because they come from afar, by the former because they are seen from afar. The wood statue on the altar is never reverenced by him who knew it as a tree trunk in the garden.

学生
我想留在中国，
但是没钱，
请大家帮忙。
愿意结婚的有机会出国

找到你合适的位置

要根据你的优点，而不是自以为是，来找到你合适的位置。要想受人尊敬，真正的捷径是通过你的优点，如果辅之以勤奋的话，这条路就会变得更短。光有诚实正直的品格是不够的，拼命争取和固步自封都是可耻的，因为借助这些手段所成就的事情被严重玷污了，以至于名声被耻辱所毁。正途是中间路线，它介于应得和进取之间。

FIND YOUR PROPER PLACE BY MERIT, NOT BY PRESUMPTION.

The true road to respect is through merit, and if industry accompanies merit the path becomes shorter. Integrity alone is not sufficient, push and insistence is degrading, for things that arrive by that means are so sullied that the discredit destroys reputation. The true way is the middle one, halfway between deserving a place and pushing oneself into it.

1
2
3

留点东西给渴望

这样你就不会因为太过幸福而痛苦。身体离不开呼吸,灵魂少不了渴求。如果一个人拥有一切，这一切就会让人幻灭，令人不满。即使是在知识领域,也应该有一些东西留待我们去了解,为的是唤醒好奇,激发希望。幸福的过度是致命的。在施以援助的时候，不完全满足是一项策略。如果不给渴望留下点什么,那就给忧惧留下了一切——这是幸福中的一种不幸状态。当渴望死去的时候,忧惧就诞生了。

LEAVE SOMETHING TO WISH FOR.

That way you will not be miserable from too much happiness. The body must respire and the soul aspire. If one possessed all, all would be disillusion and discontent. Even in knowledge there should be always something left to know in order to arouse curiosity and excite hope. Surfeit of happiness are fatal. In giving assistance it is a piece of policy not to satisfy entirely. If there is nothing left to desire, there is everything to fear — an unhappy state of happiness. When desire dies, fear is born.

图书在版编目(CIP)数据

智慧书·银色锦囊/(西)葛拉西安著;秦传安译;卜桦绘图.—北京:北京图书馆出版社,2007.8

ISBN 978-7-5013-3512-1

Ⅰ.智… Ⅱ.①葛…②秦…③卜… Ⅲ.人生哲学—通俗读物 Ⅳ.B821-49

中国版本图书馆CIP数据核字(2007)第115313号

书名 智慧书·银色锦囊

著者 (西)葛拉西安 著 秦传安 译 卜桦 绘图

出版 北京图书馆出版社(100034 北京市西城区文津街7号)

发行 010-66139745,66175620,66126153

66174391(传真),66126156(门市部)

E-mail cbs@nlc.gov.cn(投稿) btsfxb@nlc.gov.cn(邮购)

Website www.nlcpress.com

经销 新华书店

印刷 北京联兴盛业印刷有限公司

开本 720×980(毫米) 1/20

印张 10

版次 2007年8月第1版 2007年8月第1次印刷

字数 60(千字)

书号 ISBN 978-7-5013-3512-1/B·117

定价 25.00元